Exploring the World of Robot Manipulators

Edited by Serdar Küçük

Published in London, United Kingdom

Exploring the World of Robot Manipulators
http://dx.doi.org/10.5772/intechopen.111040
Edited by Serdar Küçük

Contributors
Abdelhamid Ghoul, Ahmad Shah Hizam Md Yasir, Alexandra Green, Ashley Alexander, Bethany Malone, Borislav R. Yoshinov, Denesh Sooriamoorthy, Ivet B. Koleva, Joy Massouh, Julian Tan Kok Ping, Kamel Kara, Miqdad Taqi Mohamed Mushadiq, Nohaidda Sariff, Puteri Nor Aznie Fahsyar Syed Mahadzir, Radoslav R. Yoshinov, Rasoul Farahi, Selman Djeffal, Serdar Küçük, Steven Eu Kok Seng, Um Albaneen Jamali

First published in London, United Kingdom, 2024 by IntechOpen
IntechOpen is the global imprint of INTECHOPEN LIMITED, registered in England and Wales, registration number: 11086078, 167-169 Great Portland Street, London, W1W 5PF, United Kingdom

British Library Cataloguing-in-Publication Data
A catalogue record for this book is available from the British Library

Additional hard and PDF copies can be obtained from orders@intechopen.com

Exploring the World of Robot Manipulators
Edited by Serdar Küçük
p. cm.
Print ISBN 978-0-85466-785-7
Online ISBN 978-0-85466-784-0
eBook (PDF) ISBN 978-0-85466-786-4

For EU product safety concerns:
IN TECH d.o.o., Prolaz Marije Krucifikse Kozulić 3, 51000 Rijeka, Croatia,
info@intechopen.com or visit our website at intechopen.com.

Meet the editor

Serdar Küçük received a BA and MSc from Marmara University, Istanbul, Turkey, in 1995 and 1998, respectively. He received a Ph.D. from Kocaeli University, Turkey, in 2004, where he is currently a full professor in the Department of Biomedical Engineering. He has several scientific publications to his credit, including international conference papers, journal papers, books, and book chapters. He serves as a reviewer of several well-known robotic journals. He is also an editor of scientific books. His research interests include optimization, control, and kinematics and dynamics modeling of serial and parallel robotic manipulators. Lately, Dr. Küçük has also been interested in designing electrically controlled, above-knee prosthetics and hand–wrist rehabilitation robots, surgical robots, and biomedical robotic devices. Dr. Küçük is currently at the University of South Florida in the United States of America, as part of TUBITAK 2219, to conduct research on prosthetic testing platforms.

Contents

Preface

With the development of robotic devices and their entry into our lives, interest in robotic engineering is increasing day by day. Especially in recent years, multi-robot systems (such as serial robots, parallel robots, mobile robots, and hybrid robots) have been followed with curiosity by many people. Multi-robot systems are used in many areas from surgical operations in hospitals (Da Vinci Robotic Surgical Systems) to carrying loads in mountainous terrain. The important reasons behind the preference for multi-robot systems involve better load-carrying capacity, high rigidity, better accuracy, and higher speed and acceleration. The design of multi-robot systems that perform such useful tasks involves more difficulties than the design of other engineering devices. Despite these difficulties, scientists' interest in multi-robot systems continues unabated.

Robotics engineering is at a dazzling stage when industrial robots, military robots, medical robots, rehabilitation robots, and micro-robots are considered. As a result of the intensive work of scientists and industrial robot designers, a new multi-robot design is introduced to interested people every day.

This book, *Exploring the World of Robot Manipulators*, aims to describe the latest developments in multi-robot systems in understandable language. The new book also focuses on current robotic systems and applications. I would like to thank all authors who contributed to this book with their valuable novel ideas and their knowledge of current developments in robotics.

Serdar Küçük, Ph.D.
Full Professor,
Technology Faculty,
Department of Biomedical Engineering,
University of Kocaeli,
Kocaeli, Turkey

Section 1

Medical Robotics in Healthcare

Chapter 1

Introductory Chapter: Are Robots about to Change the Way We Live?

Serdar Küçük

1. Introduction

The robot industry continues to expand and grow at full speed. While initially only industrial robots were produced, today other robot types, such as medical robots, military robots, service robots, and mobile robots, are being produced at full speed. New types of robots are produced, and new robot designs continue to surprise people more than the previous one. New types of robots are produced and new robot designs, such as the robot dog spot [1], the robotic jumping kangaroo [2], the smart robotic bird [3], and the robot Cheetah 3 [4], continue to surprise people more than the previous one. Industrial robots successfully perform important tasks such as painting, cutting, welding, pick-and-drop operations and component placement, especially in factories. Medical robots, which constitute an important group of these robots, perform a wide range of functions. Rehabilitation robots, such as Lokohelp [5] and Lokomat [6], Esa human arm exoskeleton [7] and Sarcos Master Arm [8], solve people's problems, such as a doctor. Advanced prostheses [9, 10] can work so efficiently that amputees almost do not feel the absence of their feet and hands. Surgical robots [11] can perform surgeries more carefully than surgeons.

Behind this spectacular growth of the robotics industry is not only the insatiable curiosity of scientists. At the same time, all kinds of needs of people for robots (industrial, corporate, individual) help the growth of this field. The most important reason behind this fact is that transportation of intermediate qualified workforce is becoming more difficult day by day. In other words, the world is not as generous as it used to be in training immediately qualified technicians. More importantly, the new generation, Generation Z, prefers to do work over the internet from where they sit, instead of doing repetitive, boring, and physically heavy work in factories or being a technician. This problem continues to increase day by day. Robot designers and manufacturers prefer to produce new robots that will do these jobs that the new generation rejects. In other words, robots have become an inevitable part of closing the workforce gap. In the chapters of this new book, we witness that scientists present new robotic approaches to benefit human life. While robots are doing work in factories, hospitals, schools, parks, gardens, and homes, in short, everywhere, lifestyles have slowly begun to change in every country of the world. So, what will their new lifestyle be like? Is the world ready for this change?

2. Emerging technologies and impacts on societies

There is no doubt that automation is having deep effects on labor markets, workers, and jobs. Plenty of occupations and jobs have the risk of automation [12]. Karl Marx says that technological development negatively affects the class struggle. For Marx, labor is the most important value and nothing else can replace it. According to Marx, the reason why employers invest in technology is their desire to break strikes and control the labor market [13]. Industrial robots and artificial intelligence applications have significant benefits for social, economic, and health areas [14]. One of the crucial questions is how effects industrial robots and AI applications on societal security. There is an open challenge about polarization, bias, inequalities, and vulnerable groups [15]. The rapid development of industrial robots and artificial intelligence technology applications provide enterprises with a "creative destruction" environment and both play a positive role in environmental governance and energy utilization [16, 17]. On the other hand, industrial robots, which use routine tasks in the manufacturing field, are replacing middle-skilled jobs. Especially, using industrial robots in labor-intensive manufacturing fields with lower levels of digital economy development has an important role in enhancing green innovation and intense marketing competition [18]. According to Shen and Zhang [19], artificial intelligence development in China has many benefits. These include stabilizing employment, improving traditional and physical infrastructure, strengthening knowledge management, and accelerating intelligence. According to Küçük [20], there are three main reasons for the widespread use of artificial intelligence. The human brain has continued to shrink since 1995, and the COVID-19 pandemic has caused significant memory loss and shrinkage of the human brain. Artificial intelligence ensures the security of logical data by solving complex algorithmic problems. In recent years, digital marketing platforms have increased global consumption after COVID 19, as time-consuming and complex algorithmic problems have been solved more reliably and faster with artificial intelligence. Thus, it is possible to sustain capitalism through creative destruction. On the other hand, shopping centers (street bazaars, grocery stores...) that involve face-to-face communication are decreasing, and local indigenous production is suffering [21–23]. In summary, while the widespread use of artificial intelligence hardware provides efficiency in production, there is no proven information that it reduces production costs. On the other hand, artificial intelligence hardware is spreading incredibly rapidly in communication and information technologies and supports the market economy by spreading the postmodern consumption culture to a more widespread level with its submental messages. Another reason for the widespread of artificial intelligence is the trade war between the USA and China, for the race for foreign markets. At the same time, automation and emerging technologies negatively affect middle-skilled jobs and workers. Lifelong learning, continuously improving skills, problem-solving abilities, emotional intelligence, and communication skills are necessary for high-skilled workers. A human capital that competes without stopping and resting can bring huge social problems.

3. Conclusions

As the world rapidly becomes digital, new robot designs continue to surprise more and more people. The increasing production of industrial robots, medical robots, service robots, and other robots every day causes the costs of these robots to decrease.

As a result of the decrease in robot costs, it is inevitable that robots will take over people's jobs in every field. In addition, the fact that Generation Z finds all the work in factories and other environments boring and does not want to work eliminates all obstacles to robots working in these environments. Although robots have rapidly entered all areas of business, the social consequences of this situation are not yet fully understood. If humanity will work less in the future, what will humanity do with its free time? It is a matter of curiosity what the answer to this question will be. I also believe that social scientists should think about this question from now on.

Author details

Serdar Küçük
Technology Faculty, Kocaeli University, Biomedical Engineering, Umuttepe Kampus, Kocaeli, Türkiye

*Address all correspondence to: skucuk@kocaeli.edu.tr

References

[1] The Robot Dog Spot. United States: Boston Dynamics; 2016

[2] Kangaroo B. Festo. Ruiter Straße 82 73734 Esslingen Germany; 2014

[3] Smart Bird Animation. Festo. 2011

[4] Bledt GP, Katz M, Carlo B, Wensing J, Kim P. MIT Cheetah 3: Design and Control of a Robust, Dynamic Quadruped Robot. 2018

[5] Freivogel S, Mehrholz J, Husak-Sotomayor T, Schmalohr D. Gaittraining with the newly developed "Loko help"-system is feasible for non-ambulatory patients after stroke, spinal cord and brain injury. A feasibility study. Brain Injury. 2008;**22**(7-8):625-632

[6] Colombo G, Joerg M, Schreier R, Dietz V. Treadmill training of paraplegic patients using a robotic orthosis. Journal of Rehabilitation Research and Development. 2000;**37**(6):693-700

[7] Schiele A, Visentin G. The ESA human arm exoskeleton for spacerobotics telepresence. In: International Symposium on Artificial Intelligence, Robotics and Automation in Space (iSAIRAS). Vol. 7. Nara, Japan; 2003

[8] Mistry M, Mohajerian P, Schaal S. An exoskeleton robot for human arm movement study. In: 2005 IEEE/RSJ International Conference on Intelligent Robots and Systems (IROS 2005). 2005

[9] Lee JT, Bartlett HL, Goldfarb M. Design of a semi powered stance-control swing-assist transfemoral prosthesis. IEEE/ASME Transactions on Mechatronics. 2019;**25**(1):175-184

[10] Ege M. S Kucuk Energy Minimization of New Robotic-Type Above-Knee Prosthesis for Higher Battery Lifetime. Applied Sciences. 2023;**13**(6):3868

[11] Dwivedi J, Mahgoub I. Robotic surgery - a review on recent advances in surgical robotic systems. In: Florida Conference on Recent Advances in Robotics, 2012. Boca Raton, Florida; 2012

[12] Makridis CA, Mishra S. Artificial intelligence as a service, economic growth, and well-being. Journal of Service Research. 2022;**25**(4):505-520. DOI: 10.1177/10946705221120218

[13] Küçük Ö. Kapitalizm ve Tüketim Temelinde Yükselen Tüketici Hakları. Anemon Muş Alparslan Üniversitesi Sosyal Bilimler Dergisi. 2020;**8**(6):1763-1777. DOI: 10.18506/anemon.766277

[14] Taylor SM, Gulson KN, McDuie-Ra D. Artificial intelligence from colonial India: Race, statistics, and facial recognition in the global south. Science, Technology, & Human Values. 2023;**48**(3):663-689. DOI: 10.1177/01622439211060839

[15] Liang P, Sun X, Qi L. Does artificial intelligence technology enhance green transformation of enterprises: Based on green innovation perspective. Environment, Development and Sustainability. 2023. DOI: 10.1007/s10668-023-04225-6

[16] Chhillar D, Aguilera RV. An eye for artificial intelligence: Insights into the governance of artificial intelligence and vision for future research. Business & Society. 2022;**61**(5):1197-1241. DOI: 10.1177/00076503221080959

[17] Küçük Ö. Dünya Bankası Ve Sürdürülebilir Kalkınma: Dijital

Platformlar Ve Sürdürülebilir Turizm. Uluslararası Beşeri Bilimler Ve Eğitim Dergisi. 2022;**8**(17):120-152

[18] Küçük Ö. Endüstri Ötesi Toplum, Popüler Kültür, Sürdürülebilir Turizm ve Yapay Zekâ. In: Deniz Beste Çevik Kılıç, editor. Güzel Sanatlar Alanında Uluslararası Teori, Araştırma ve Derlemeler kitabında. Ankara: Serüven Yayınevi; 2021. Available from: https://www.seruvenyayinevi.com/Webkontrol/uploads/Fck/gsfekim2023_1.pdf

[19] Shen Y, Zhang X. The impact of artificial intelligence on employment: The role of virtual agglomeration. Humanities and Social Sciences Communications. 2024;**11**:122. DOI: 10.1057/s41599-024-02647-9

[20] Küçük Ö. Endüstri Ötesi Toplum, Popüler Kültür, Sürdürülebilir Turizm ve Yapay Zekâ. In: Deniz Beste Çevik Kılıç, editor. Güzel Sanatlar Alanında Uluslararası Teori, Araştırma ve Derlemeler kitabında. Ankara: Serüven Yayınevi; 2023. Available from: https://www.seruvenyayinevi.com/Webkontrol/uploads/Fck/gsfekim2023_1.pdf

[21] Cheong PH. Bounded religious automation at work: Communicating human Authority in Artificial Intelligence Networks. Journal of Communication Inquiry. 2021;**45**(1):5-23. DOI: 10.1177/0196859920977133

[22] Küçük Ö. Dünya Bankası ile Yeni Bir Sosyal Politika Öznesi Olarak Beşeri Sermaye. Gaziantep University Journal of Social Sciences. 2020;**19**(2):475-498. DOI: 10.21547/jss.686386

[23] Stahl BC, Antoniou J, Bhalla N, et al. A systematic review of artificial intelligence impact assessments. Artificial Intelligence Review. 2023;**56**:12799-12831. DOI: 10.1007/s10462-023-10420-8

Chapter 2

An Overview of Rehabilitation Robots and their Application

Rasoul Farahi

Abstract

In the community health system, non-pharmacological treatments are at the forefront of countries' treatment. One of these treatments is rehabilitation. Rehabilitation is considered a very broad scientific field in the set of health and medical services and helps people recover after suffering from problems such as stroke, spinal cord injuries, orthopedic surgery, concussions, burns, hearing loss, central auditory processing disorder, balance problems, overcome their problem as much as possible and regain their previous functional independence. This process is expensive and requires a lot of time and patience. There are many exercise machines for rehabilitation purposes. These types of machines are not flexible during operation, and this lack of flexibility destroys their ability to face unwanted conditions during physiotherapy. Therefore, the necessity of using mechatronic technologies in the field of rehabilitation is clearly evident. In this chapter, we will first give an overview of a number of research projects carried out in the field of rehabilitation robots (mainly lower body) and, since our main focus is more on the optimal control of the intelligent system (mechatronics), in the continuation of the strategies we describe different controls applied to rehabilitation robots and examine their strengths and weaknesses.

Keywords: rehabilitation, robot, robotic rehabilitation, mechatronics, mainly lower body

1. Introduction

The human movement system since birth is an extremely harmonic, regulated and purposeful system that has always attracted the attention of scientists interested in physiotherapy until 1945 due to the absence of physiotherapists and the existence of only physiotherapy technicians and activities in the rehabilitation and rehabilitation departments, of a part of It has been rehabilitation, but after 1945 and the emergence of the modern physiotherapist, it included treatment and rehabilitation, and after 2000, with professional doctorate titles, physiotherapy included prevention, treatment and rehabilitation.

In the last decade, the introduction and use of robotic technologies for the purpose of rehabilitation has turned from concept to reality. In general, rehabilitation robots are classified on three basis: one is based on the therapeutic field, the second is based on how to interact with the patient, and the third is based on the purpose of use in the therapeutic field. Rehabilitation robots are divided into upper and lower body groups. Upper body robots are generally used to restore the mobility of the arm or forearm or fingers. On the other hand, lower body robots have been used to restore the mobility of the knee, thigh, hip, ankle, and foot. The robots that are made for knee treatment are also lower body robots, but the general purpose of lower body robots is mostly a robot that must maintain the balance of the patient's weight while working and the patient while working with it, is standing. The difference between upper body and lower body rehabilitation is that the weight of the body is on the lower body, and it plays a major role in losing the balance of the lower body. In lower body robots where both legs work together, coordination is very important. Upper-body robots benefit from virtual reality technology for their work, while this issue is rarely seen for lower-body robots [1, 2].

In another classification, rehabilitation robots are divided into two groups: external skeleton and final effector. The sacral skeleton is a robot that is made in the shape of one of the body parts and is connected to the disabled part at several points and moves it along with it. The advantage of using the exoskeleton is that it always supports the disabled member and is more useful in obtaining movement coordination. Controlling these robots is more difficult than other types of robots due to the number of connections and the need to coordinate them with each other. In contrast to the external skeleton, the final effector must be mentioned, which is connected to the disabled member at only one point. The final effector has a much higher force interaction rate than the external skeleton, and the patient feels more free while working with it.

In the last classification, rehabilitation robots are divided into two groups: collaborators and therapists. Collaborative robots help the patient to do the tasks of daily life and there is no therapeutic motivation. But therapeutic robots are used to increase the patient's movement ability.

2. Completed projects in the field of rehabilitation robots

The designs that have been developed so far in the world for the rehabilitation of the lower limbs of physically disabled patients can be divided into four general categories in terms of how the mechanism works:

2.1 First category: body weight support system

There are mechanisms in which a body weight support system is used while walking. The structure of this system is shown in **Figure 1**.

By conducting a series of tests on disabled patients using this system, the result has been that the ability to walk has been greatly improved compared to the old methods. A test sample was conducted at the National Rehabilitation Hospital of the Netherlands in 2002 on sixty seriously injured patients. In this way, these sixty people were randomly divided into two groups. For the first group of old physiotherapy methods and the second group of new physiotherapy system (**Figure 1**) called ZeroG was used. Both groups were treated for 2 hours a day, 5 days a week, for 14 weeks, and after this period, the improvement of walking on different and even heterogeneous surfaces was visible for the second group much more than the first group [3–6].

2.2 Second category: treadmill

There are mechanisms in which treadmills are used. These devices are only able to simulate human walking on smooth surfaces. Based on the conducted research, walking on a treadmill alone cannot have a great effect on the rehabilitation

DOI: http://dx.doi.org/10.5772/intechopen.1004346

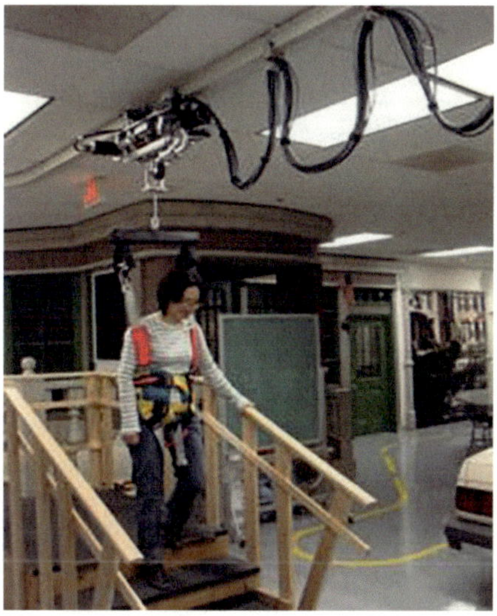

Figure 1.
Walking on the ground using a body weight support system.

of stroke or spinal cord injury patients; Because it does not have the ability to simulate many situations, such as climbing the stairs, rotating the ankle joint, etc. (these capabilities are available in the first type of systems). Accordingly, in addition to the treadmill, other side mechanisms are also used in these designs [7, 8]. One of the most famous designs in this field is the Lokomat robotic set, which is shown in **Figure 2**.

This device is made by the Swiss Hocoma® company and is commercially available on the market today. In this design, the movement of each of the hip, knee and ankle joints is controlled by a separate electric actuator. Also, a degree of freedom for the movement of the upper body in the vertical direction is considered, for which the necessary force is provided by a pneumatic operator, and the amount of this force can be adjusted by an active control system. Also, the device includes force and position sensors in several places, as well as a display for simulating virtual environments. The amount of interaction forces is also shown to the patient by the displays [2].

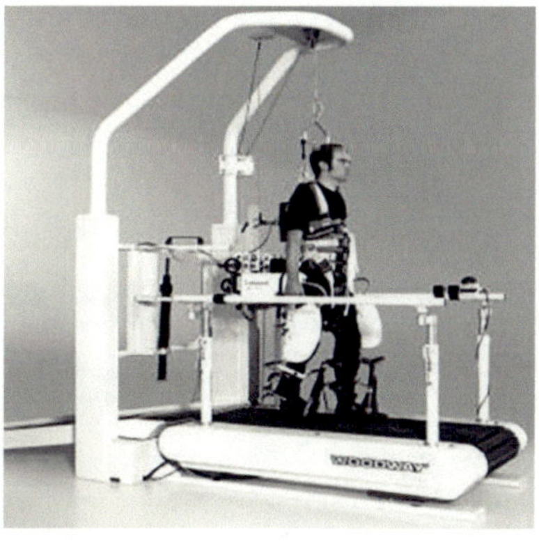

Figure 2.
Lokomat robotic assembly.

2.3 The third category: exoskeleton

There are mechanisms in which external skeletonization is used. The skeleton is connected to the patient's body from several points and is usually made of several pneumatic actuators that coordinately help the disabled person's muscles and create natural movements. One of these simple designs is a device called CPMs(Continue Passive Motions) and is shown in **Figure 3**. This device is used in most pharmaceutical centers for the treatment of disabled patients and rehabilitation purposes. The concept of CPM was first introduced in the 1970s by Salter and Simmonds.

In some cases, the CPM device is not suitable for the physiotherapy treatment process. During rehabilitation, patients sometimes move their leg beyond expected limits as a reaction to an operation, quite by accident. Old machines such as CPM are not able to respond in these conditions. If a reflex causes the patient's leg to move beyond the expected limit (the range of motion of the machine), an inappropriate load is produced and can damage the patient's muscles or tendon tissues [9].

Figure 3 shows the plan of the University of Michigan researchers. This design is used to rehabilitate muscles effective in ankle movement, which is made of two pneumatic actuators. The noteworthy point in this plan is the use of EMG(Electromyography) signals to help stimulate the patient's muscles along with the mechanical system [1, 2, 6].

Another type of these designs is RGO, which was developed by a research group at Stanford University, England, and is shown in **Figure 4**. In this mechanism, the

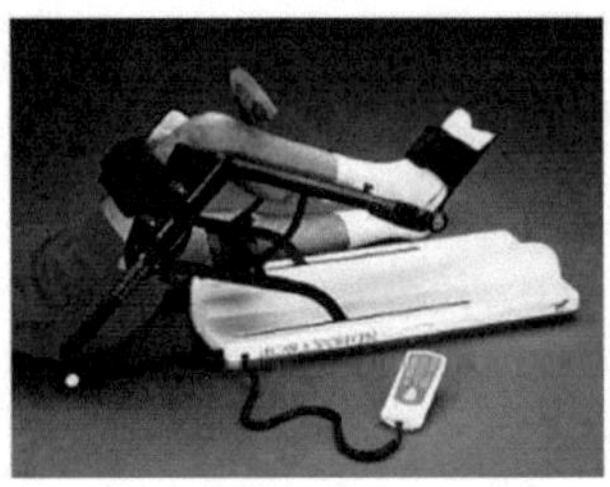

Figure 3.
CPM for lower limbs.

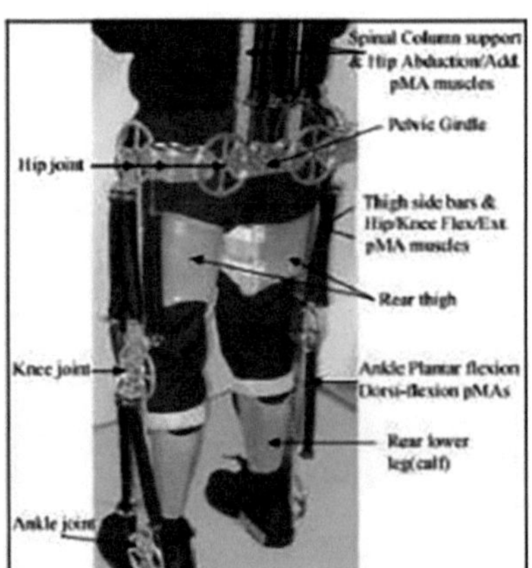

Figure 4.
Image of RGO.

movement of each leg has 5 degrees of freedom, of which three degrees of freedom are related to the hip area, one degree of freedom for knee movement and one degree of freedom for wrist movement. The accuracy of changing locations in this system is 1% and its bandwidth is 5 Hz.

The ALEX robotic set is another type of these designs that was made at the University of Delaware. **Figure 5** shows the image and components of this device [8]. In addition to the degrees of freedom related to movement in the sagittal plane, this device has the ability to create horizontal movement in the frontal plane, the ability to rotate the upper body around the vertical axis, and the ability to create adduction and abduction movements in the hip joint [6].

Another one of these plans is the LOPES robotic set, which is shown in **Figure 6**. This robot was made at the University of Twente in the Netherlands and has the ability to move the whole body in front-back and left-right directions. The weight is passively restrained by a four-bar mechanism.

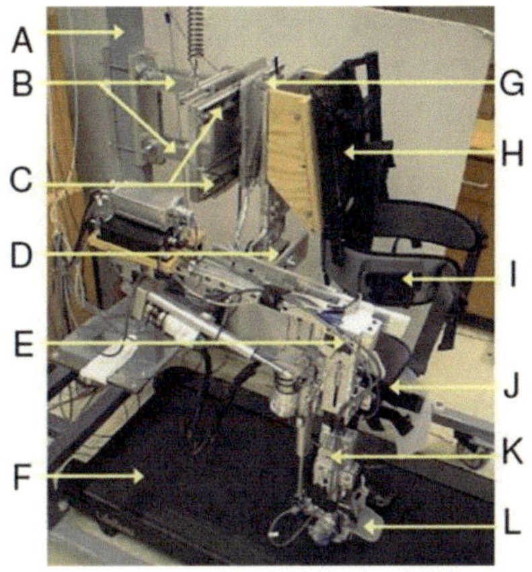

Figure 5.
Components of ALEX rehabilitation device: A: Vertical displacement mechanism; B: Four-bar mechanism for vertical movement; C: Sliding mechanism for horizontal movement in the frontal plane; D: Hinged connection to create abduction and adduction movements of the leg; E: Hip joint to create thigh movement in the sagittal plane; F: Treadmill; G: Hinged connection for turning around the vertical axis; H: Person support from the back, I: Upper body support harnesses, J: Thigh support, K: Knee joint, L: Foot entry point.

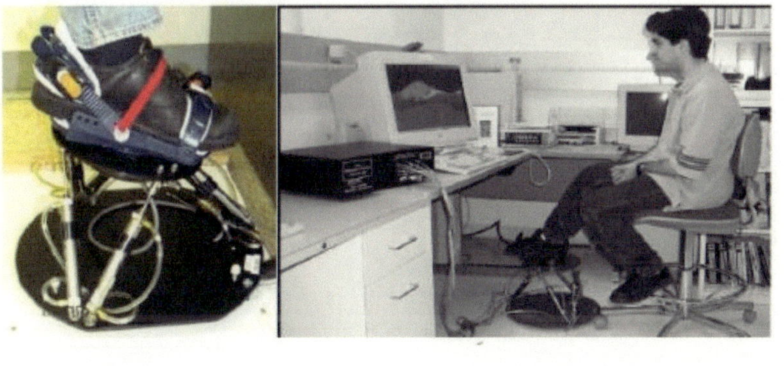

Figure 6.
LOPES robotic set.

2.4 The fourth category: moving pages

There are mechanisms in which moving plates are used under the soles of the patient's feet in order to create the desired movement. In these mechanisms, by planning the movement of the moving plate, the whole leg or a part of it is forced to move or they are helped in creating desired movements.

The most important projects in this field are as follows:

a. **Figure** 7 shows the plan of Rutgers University researchers. This mechanism is used to rehabilitate stroke patients who have disabilities in the ankle area. In this design, a parallel mechanism is used to create different movements. In addition to using the virtual reality environment with the help of a computer, the quality of the rehabilitation process has been improved [10].

b. The research group of Rutgers University has presented a plan in which two parallel mechanisms are used under both legs of the patient, with the difference that a person uses this robotic set while standing. This robot is called Mobility Simulator, and it is shown in **Figure 8**. In this design, a part of the person's weight is kept by straps tied to him from above.

In the following, we will have an overview of some other completed projects in the field of rehabilitation robots.

In 1998, a system with two robots was introduced, each of which included 2 degrees of freedom. This system was suitable for minor unpredictable movements (low range of motion and lower limbs) and was also used to predict the parameters of human body parts. In 1990, a robotic system for rehabilitation with more unwanted postures (more range of motion) was developed. In 1995, a collaborative arm rehabilitation system was introduced. Another system for upper limb rehabilitation that included 5 degrees of freedom robotic

Figure 7.
Rutgers university plan for rehabilitation of ankle muscles.

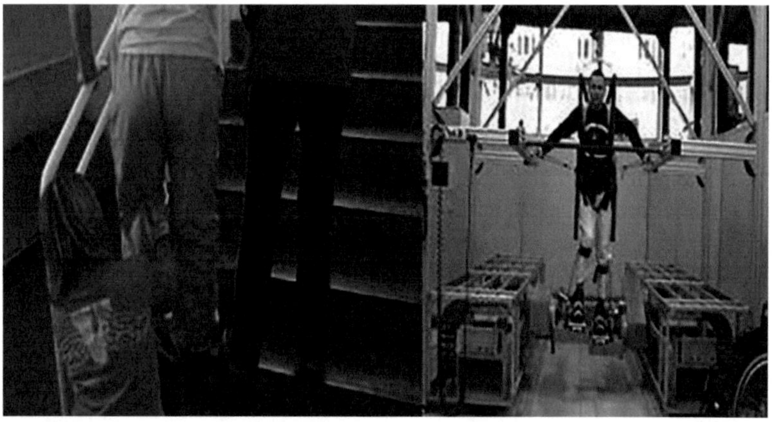

Figure 8.
A picture of the mobility simulator robot.

arm was developed in 1997 under the name of the MULOS project. In 1998, a system named MIT-MANUS was presented, which included a clinical robot with a neurological rehabilitation system. This system included a robot with multiple degrees of freedom and suitable for injured and traumatic patients. Also, in 1990, another system using the Puma 240 robot was introduced, which was used for active and passive rehabilitation of the upper limbs. In 1999, a 3-degree-of-freedom pneumatic system was developed for rehabilitation, which uses PD and impedance controllers. In 2000, another work for rehabilitation was the REHABROB project, which involved two industrial robots. The main idea is based on the required force and position provided by sensors placed on different parts of the patient's body during the rehabilitation process. Then, the industrial robots repeat the same procedure based on the information collected from the sensors and the requested force and position. These cases and other projects done in this field can be found in Refs. [9–11].

3. Parallel robots

Considering that the motivation of robotics science and technology is the development of mechanical systems in which robots do the work instead of humans, it is almost obvious that most of the attention will be directed towards open series rings as a robot arm. Such robotic arms have advantages such as a large working space and high maneuverability, similar to a human arm, but usually due to a structure similar to a sniper's arrow, they have a small load carrying capacity. As a result, on the one hand, they suffer a lot of rise and, on the other hand, they vibrate at high speeds. Also, as mentioned, they have a large working space, but they do not have the ability to accurately position themselves. In short, open arms have the advantages and disadvantages of the human arm. Therefore, in practice, when the ability to carry heavy loads, good dynamic efficiency and accurate positioning are desired, it is better to use a method other than traditional series arms. In general, it can be concluded that robot arms whose final actuators are connected to the ground through closed rings with parallel actuators have higher rigidity and accuracy [3]. For this reason and because of the controllable degrees of freedom, parallel mechanisms have been given more attention for some applications, and since 1980, a relatively large amount of research has been done in this field. One of these mechanisms is the Stewart mechanism, which we will describe below:

3.1 Stewart mechanism

This six-degree-of-freedom mechanism was first used by D. Stewart in 1965 in a flight simulator. The Stewart mechanism used in 1965 consisted of a triangular plate supported by spherical joints on three legs of variable length. This mechanism also had angular height gauges that were connected to the ground by two simple joints. After that, the mechanism proposed by Goff, which was used in the car tire tester and used six linear operators, turned Stewart's mechanism into a completely parallel mechanism. Then Hunt proposed in 1978 that parallel mechanisms such as the Stewart mechanism, which is used in flight simulations, could be used as a robotic arm. These mechanisms have special advantages, such as higher rigidity and more accurate positioning than conventional robots [4].

Today, the Stewart mechanism consists of two solid bodies (one base and the other moving plate) which are connected through six linear actuators. These operators are connected to the plates in two ways. One is to be connected to the plates from both sides with spherical joints, and the other is to be connected to the plates from one side connected to plates with spherical joints and joints the other hand with universal, joint. The first mode is called 6-SPS and the second mode is called 6-UPS. In cases where there is no emphasis on a specific type, any of the above situations can be considered. Of course, we note that these mechanisms have a similar dynamic behavior and the only difference is that the SPS mode has six additional degrees of freedom related to the rotation of the actuators around its axis, which does not affect

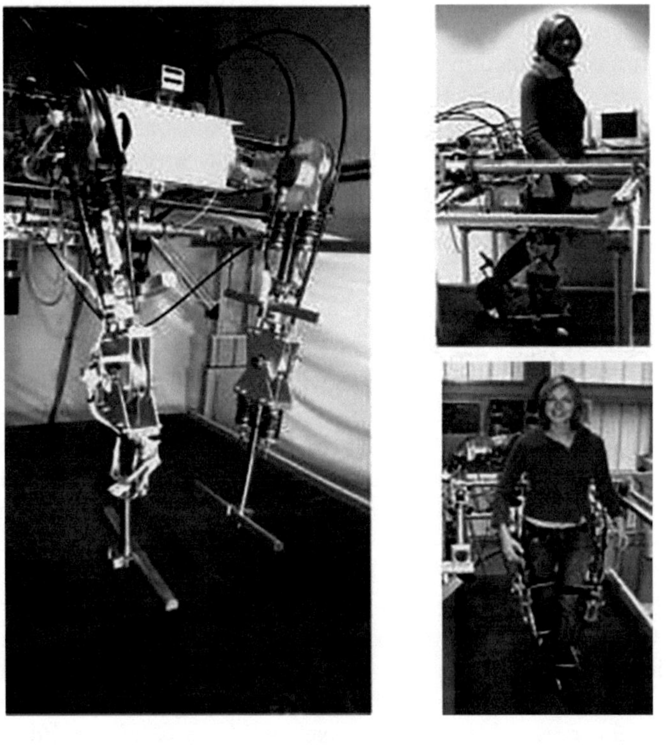

Figure 9.
Parallel robot mechanism.

the overall performance of the mechanism [4, 5]. Due to their closed nature, parallel mechanisms also have a number of inactive joints, which makes all the joints of this mechanism not controllable independently [1, 2, 6]. In fact, if the mechanism does not have additional stimulation, the number of independently controlled joints is equal to the number of degrees of freedom of the moving plate of the mechanism (**Figure 9**).

Flight simulators can be mentioned among the oldest applications of Stewart robots. Examples of its applications are shown in the following **Figures 10–14**.

The following **Figures 15–18** show other examples of applications of parallel robots in industries:

Figure 10.
Parallel robot mechanism.

Figure 11.
Stewart flight simulator.

Figure 12.
Airbus A320 flight simulator.

Figure 13.
The largest driving simulator built at the University of Iowa.

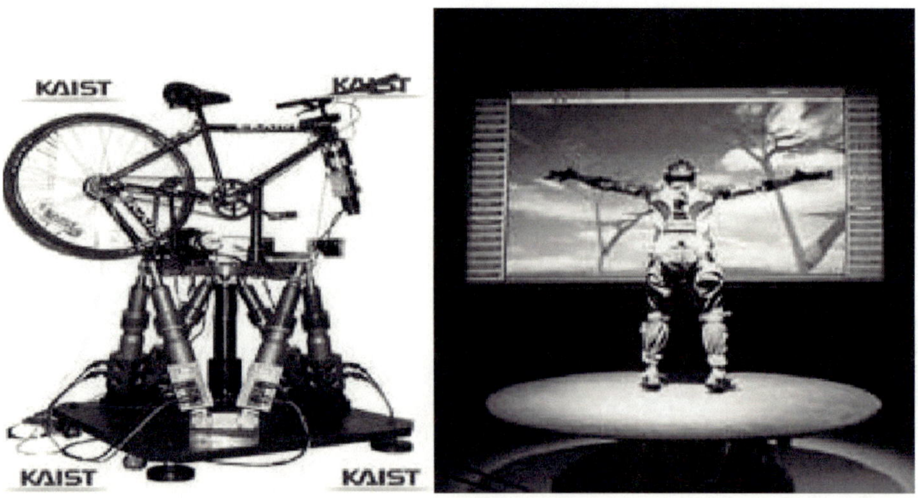

Figure 14.
KAIST and Motek simulators.

Figure 15.
Paris positioning robot.

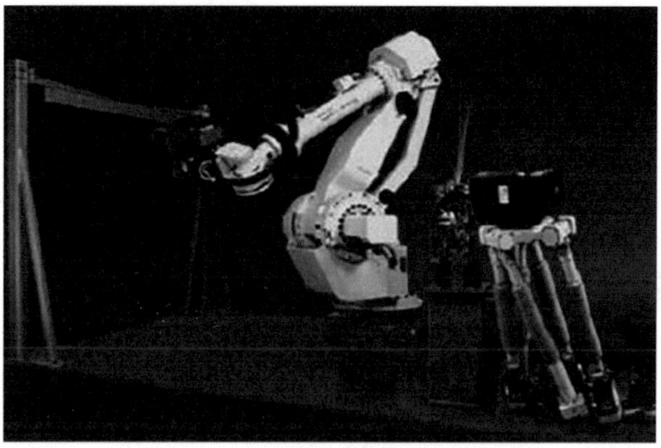

Figure 16.
An example of the industrial application of the F-200i robot in maintaining and transporting the cylinder block.

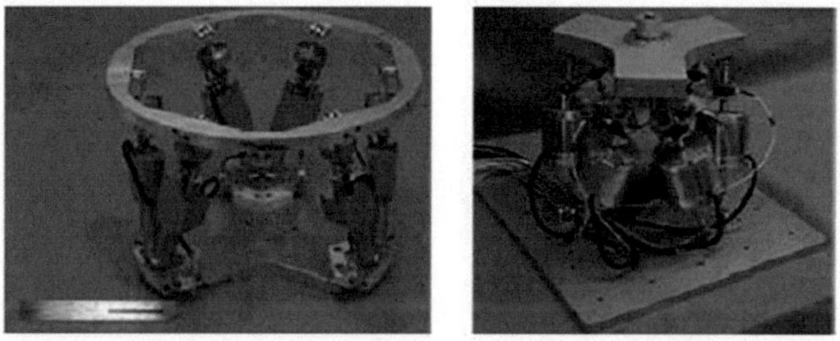

Figure 17.
PH1 and PHEX1 vibration isolators in CSA engineering group.

Figure 18.
Mars robot with 6-ups structure as a surgeon's help.

4. Conclusions

Robotic rehabilitation is a field of research dedicated to understanding the enhancement of rehabilitation through the use of robotic physiotherapy devices. Robotic rehabilitation involves the development of robotic devices to assist with various sensory and motor functions such as arms, hands, feet, ankles). Advancement and development of various programs to help therapeutic sports and evaluate the performance of the motor system of the patient (patient movement, where robots are used instead of aids as a sensory aid for treatment) Rehabilitation using robots is more tolerable for the patient, and it has also been determined that it is an effective treatment method for patients who suffer from movement problems, such as people with a stroke. Therefore, in this chapter, lower trunk rehabilitation robots and their use in limb rehabilitation will be examined. We discussed the patient's lower body in four categories: 1- Maintaining body weight, 2- Treadmill, exoskeleton (external skeleton) and 4- Moving plate. The results of investigations and research show that the structure of parallel robots whose final executor through Closed loops are connected to the ground with parallel operators and have higher precision and accuracy for rehabilitation applications.

Author details

Rasoul Farahi
Department of Computer Engineering, Islamic Azad University, Mahabad Branch, Mahabad, Iran

*Address all correspondence to: stu.rasoul-farahi@iaut.ac.ir

References

[1] Aminiazar W, Najafi F, Nekoui MA. Optimized intelligent control of a 2-degree of freedom robot for rehabilitation of lower limbs using neural network and genetic algorithm. Journal of Neuroengineering and Rehabilitation. 2013;**10**(1):1-11

[2] Azar WA, Nazar PS. An optimized and chaotic intelligent system for a 3DOF rehabilitation robot for lower limbs based on neural network and genetic algorithm. Biomedical Signal Processing and Control. 2021;**69**:102864

[3] Dong M, Zhou Y, Li J, Rong X, Fan W, Zhou X, et al. State of the art in parallel ankle rehabilitation robot: A systematic review. Journal of NeuroEngineering and Rehabilitation. 2021;**18**(1):1-5

[4] Dıaz I, Gil JJ, Sánchez E. Lower-limb robotic rehabilitation: Literature review and challenges. Journal of Robotics. 2011;**2011**:1-1

[5] Belagaje SR. Stroke rehabilitation. CONTINUUM: Lifelong Learning in Neurology. 2017;**23**(1):238-253

[6] Amini Azar W, Akbarimajd A, Parvari E. Intelligent control method of a 6-DOF parallel robot used for rehabilitation treatment in lower limbs. Automatika: časopis za automatiku, mjerenje, elektroniku, računarstvo i komunikacije. 2016;**57**(2):466-476

[7] Bernhardt M, Frey M, Colombo G, Riener R. Hybrid force-position control yields cooperative behavior of the rehabilitation robot LOKOMAT. In: 9th International Conference on Rehabilitation Robotics, ICORR2005. IEEE: CPM; 28 Jun 2005. pp. 536-539

[8] Qassim HM, Wan Hasan WZ. A review on upper limb rehabilitation robots. Applied Sciences. 2020;**10**(19):6976

[9] Shi D, Zhang W, Zhang W, Ding X. A review on lower limb rehabilitation exoskeleton robots. Chinese Journal of Mechanical Engineering. 2019;**32**(1):1-11

[10] Nordin N, Xie SQ, Wünsche B. Assessment of movement quality in robot-assisted upper limb rehabilitation after stroke: A review. Journal of Neuroengineering and Rehabilitation. 2014;**11**(1):1-23

[11] Dawson-Elli AR, Adamczyk PG. Design and validation of a lower-limb haptic rehabilitation robot. IEEE Transactions on Neural Systems and Rehabilitation Engineering. 2020;**28**(7):1584-1594

Chapter 3

Robotic-Powered Prosthesis: A Review and Directions

Nohaidda Sariff, Denesh Sooriamoorthy, Ahmad Shah Hizam Md Yasir, Puteri Nor Aznie Fahsyar Syed Mahadzir, Joy Massouh, Miqdad Taqi Mohamed Mushadiq, Julian Tan Kok Ping and Steven Eu Kok Seng

Abstract

Robotic prostheses involve the utilization of artificial limbs designed for optimal power efficiency, significantly enhancing users' mobility and independence. The primary focus in prosthesis development is on aspects related to power efficiency, aiming to create more advanced and energy-efficient solutions in the future. The initial discussion will delve into the state-of-the-art advancements in prosthesis robotics. Issues and challenges associated with robotic-powered prostheses, such as limited battery lifespan and power-to-weight balance concerns, will be explored. Recent approaches incorporating energy-efficient design strategies, including regenerative systems, actuation selection, power transmission mechanisms, and material selection, will also be examined. The strengths and limitations of these approaches will be highlighted. In conclusion, the presentation will outline future directions for power prosthesis robotics, addressing gaps in the current development of this field.

Keywords: robot prostheses, robot-powered prosthesis, robot prosthesis energy, robot prosthesis limbs, prosthesis robotics

1. Introduction

Robotic prostheses refer to the use of artificial limbs that are power-efficient and can significantly enhance a user's mobility and independence [1, 2]. These prostheses can assist individuals and replace a missing limb, driven by the need for esthetic appeal, professional requirements [3], or the pursuit of personal independence, leading to an expansion in research in this field.

The development of these prostheses involves various aspects related to power efficiency [4–6], such as energy harvesting, energy storage [6], and energy management strategies [6, 7]. Power-efficient prostheses can significantly enhance the user's mobility and independence while also reducing energy consumption [1], a critical aspect considering the limited battery life of these devices. For instance, power-efficient lower limb prostheses have been used to assist individuals with amputations in performing daily activities such as walking, running, and climbing stairs with less energy expenditure [2]. Similarly, power-efficient upper limb prostheses enable users to perform tasks ranging from simple object manipulation to complex tasks like writing and eating, thereby improving their quality of life [5]. Moreover, power efficiency in robotic prostheses also contributes to the longevity and durability of the device, reducing the need for frequent replacements or repairs and lowering the overall physical and psychological burden on the user [2].

Given the significant role of power-efficient robotic prostheses in enhancing human capabilities, researchers have been actively exploring various aspects related not only to power efficiency but also to material innovation, design techniques (i.e., actuator, mechanism, and control system). This active research is expected to lead to the development of more advanced robotic prostheses in the future [8]. The main strategy is to reduce the cost of current solutions while increasing the function of the prosthesis through stronger materials, effective sensing [9], more powerful and versatile control systems, and longer use times through regenerative systems.

This paper contributes (i) a detailed analysis of fundamental elements and highlighted challenges in robotic prosthesis, and (ii) an exploration of future directions for robotic prostheses, offering recommendations and insights. The subsequent sections are organized as follows: the first section delves into the current state-of-the-art in prosthetics, the second section outlines the hurdles in prosthetic robotics, the third section examines issues related to prostheses, and the final section concludes with challenges and suggestions for future research in prosthetic robotics.

2. State-of-the-art in prosthesis robots

The development of prosthetic robots has steadily grown from the 1980s to 2024, as analyzed from SCOPUS using the keyword "prosthesis robot" for this review, as illustrated in **Figure 1**. **Figure 2** shows that the publication rate on this topic has exponentially increased from 2012 to the present for both SCOPUS and Web of Science (WoS). The largest number of publications is in robotics journals, followed

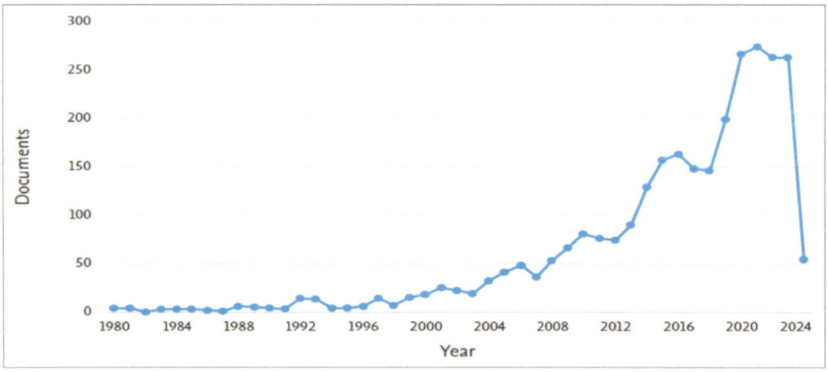

Figure 1.
The SCOPUS publication rate from 1980 to 2024.

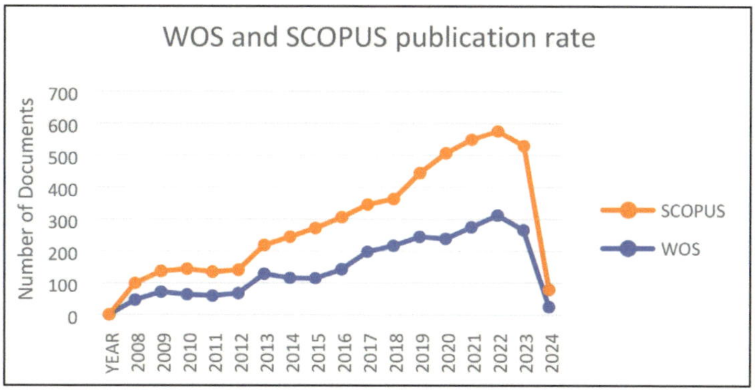

Figure 2.
The publication rates in the Web of Science (WoS) and SCOPUS from 2012 to 2024.

by biomedical engineering, electrical engineering, automation control systems, and various other engineering domains, sciences, and medical fields.

The concept of prosthetic limbs has been around for centuries, with early examples dating back to ancient Egypt. However, modern development of robotic prostheses began in the early 1980s. One of the earliest examples of a prosthesis with robotic components is the "Stanford/JPL Hand," developed in the 1980s at Stanford University and NASA's (National Aeronautics and Space Administration) Jet Propulsion Laboratory. This robotic hand was computer-controlled and capable of performing basic grasping motions.

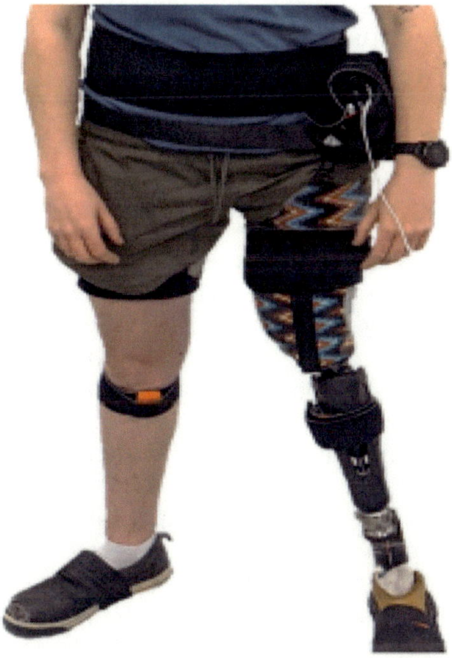

Figure 3.
A-mode ultrasound system and Xsens inertial measurement unit (IMU)-based motion capture (Mocap) system for lower limb [13].

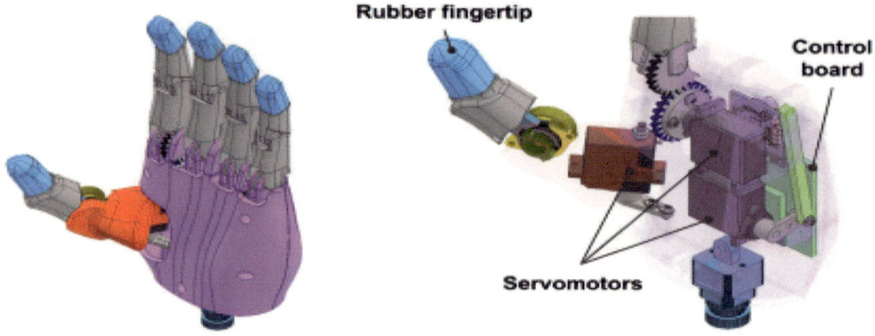

Figure 4.
Internal and external view of upper limb prosthesis [18].

In recent years, advances in materials, sensor technology, design, and control have continued to drive the development of more advanced and sophisticated prosthetic robots. The evolution shows the development from simple prosthesis devices like artificial limbs to more complex devices like prosthetic hands and bionic limbs that can be controlled by the user's muscles or nerves. Today, there are rapid advancements in prosthetic robots, including the development of brain-machine interfaces, advanced control algorithms, and advanced sensory systems, with the integration of artificial intelligence (AI) and machine learning (ML). Based on the 15 selected recent papers from Web of Science in 2024, research on upper limbs [10–12] and lower limbs [13–16] prosthetic robots was balanced at 50%. Most of them focus on improving the control system [17, 18] and design [19–24] to ensure the prostheses are more functional and natural for various applications, as shown in **Figures 3** and **4**. Despite these significant advancements, there is still room for improvement to continuously contribute to this area of prosthetic research.

3. Challenges

3.1 Power efficiency

The quest for power efficiency in robotic prostheses has led to the exploration of innovative technologies and design principles. One promising avenue is the integration of advanced energy storage solutions, such as ultra-high-efficiency batteries and energy-dense capacitors, to maximize power output while minimizing the overall weight of the prosthesis limb. The power output of a prosthesis device is a critical factor that determines its functionality and efficiency. This power output is primarily dependent on two components: the power supply and the motor [12, 25].

A prime example of a prosthesis device that utilizes these components effectively is the Power Knee by Ossur [26]. This device is specifically designed for individuals who have undergone above-knee amputations. The Power Knee is unique, in that it is motor-powered, allowing it to provide powered extension. This feature is particularly beneficial when the user is standing up, as it reduces the physical strain on the user. However, it's important to note that this powered extension requires a significant amount of power [26].

The Power Knee is considered a breakthrough in prosthesis technology as it works as an integrated extension of the user, replacing true muscle activity to bend and straighten the knee as required [25, 26]. It uses advanced algorithms to detect human movement patterns, thereby learning and adjusting to the wearer's speed and cadence in real time. This intelligent design allows the Power Knee to provide a more natural and efficient walking experience [25]. Additionally, advancements in energy harvesting techniques, including kinetic and solar energy capture, are being pursued to enable the prosthesis to generate and store its own power during use [27, 28].

Furthermore, the development of sophisticated energy management systems is essential to ensure optimal utilization of stored energy, thereby prolonging the operational duration of the prosthesis without compromising its performance. These systems involve intelligent control algorithms and feedback mechanisms to adapt to the user's movements and optimize power consumption in real time [2, 29, 30].

As the field of robotic prosthesis continues to evolve, interdisciplinary approaches that leverage expertise in materials science, robotics, and bioengineering will be instrumental in realizing the next generation of power-efficient and seamlessly integrated robotic prosthesis devices [1, 25, 26, 31].

3.2 Weight

While significant progress has been made in enhancing the power efficiency of robotic prostheses, the challenge of weight balance remains a critical aspect in the development of these devices [32]. The need to achieve a balance between power output and the weight of the prosthesis limb is essential to ensure the user's comfort and mobility [33, 34].

The weight of the robotic prosthesis directly impacts the user's ability to move freely and comfortably [28]. Excessive weight can lead to fatigue and discomfort, limiting the user's mobility and overall satisfaction with the prosthesis device. Therefore, achieving optimal weight balance is crucial for enhancing the user's experience and ensuring long-term usability of the prosthesis limb [35].

To address the issue of weight balance, researchers and engineers are exploring innovative design concepts and materials to minimize the overall weight of robotic prostheses without compromising their structural integrity and functionality [29]. Integrating lightweight yet durable materials, such as carbon fiber composites and advanced polymers, into the design of the prosthesis components is one approach being pursued to achieve a favorable balance between weight and strength [2, 36].

Furthermore, advancements in additive manufacturing techniques allow for the creation of intricate lattice structures and hollow components that reduce overall weight while maintaining structural robustness. These lightweight design approaches contribute to the overall weight optimization of robotic prostheses, enhancing the user's comfort and promoting greater mobility [30].

3.3 Power-to-weight

A significant challenge that researchers are focusing on in the field of robotic prostheses is the power-to-weight ratio [6]. This issue is complex due to the need to strike a balance between the power output and the weight of the prosthesis device.

Various researchers have proposed the development of efficient power systems [1] and lightweight materials for practical use in robotic prostheses [2]. Issues related to the weight of the battery, the power consumption of the motors, and the overall weight of the prosthesis device are among the problems that might affect the performance and usability of the device.

The power-to-weight ratio is crucial as it directly impacts the user's comfort and mobility [6]. A high-powered, heavy prosthesis may cause discomfort and fatigue, while a lightweight, low-powered/unpowered prosthesis may not provide the necessary functionality [8]. Therefore, finding the right balance is key to the successful design and implementation of robotic prostheses. Environmental sensing in robotic arms is essential for the bio-integration and training period for any amputee wishing to have a robotic prosthesis [9].

3.4 Material innovation

The use of lightweight materials is being explored to reduce the overall weight of the prosthesis device [5]. This is important as the weight of the battery [37], the power consumption of the motors, and the overall weight of the prosthesis device might affect the performance and usability of the device [26, 37, 38].

4. Prosthesis research-related

4.1 Regenerative energy system

During human locomotion, the muscles and tendons of the ankle joint collaborate to generate force and perform negative power in specific gait phases, utilizing metabolic energy [39]. This energy can be harnessed to enhance efficiency [40–42]. Some studies have explored the use of insoles to convert this mechanical energy into electrical energy, which can then be used to recharge robotic devices [43]. However, the total energy harvested from the rotation of the ankle joint or the foot touching the ground is quite small, only in the megawatt range [43].

In the realm of powered lower extremity prostheses, many studies have focused on integrating spring structures to harness energy during the early and middle stance phases and release it in the late stance phase [44, 45]. A regeneration strategy has been proposed to harness energy from electromagnetic induction due to ankle joint rotation during the stance phase for a lower extremity robotic prosthesis [46]. This involves the use of six MOSFETs (metal-oxide-semiconductor field-effect transistors) that are turned ON/OFF according to a predefined duty cycle, acting as an electromotive force (EMF) inverter. The inverted current then flows into a capacitor to enhance energy efficiency. Moreover, some studies have conducted bench tests on a powered transfemoral prosthesis to verify the feasibility of electrical regeneration [47, 48]. For a robotic prosthesis, the total power equals the battery power plus a continuous regeneration power. This basic regeneration principle is demonstrated in **Figure 5**, where the prosthesis ankle joint with damping behaviors [49] is driven due to human dynamics during the stance phase, leading to the development of an EMF according to electromagnetic induction theory.

4.1.1 Passive kinetic energy storage systems

Biomechanical research indicates that muscles expend metabolic energy to generate force and carry out negative work [39] during the initial stance phase. In-depth studies reveal that during the controlled dorsiflexion and plantar flexion stages,

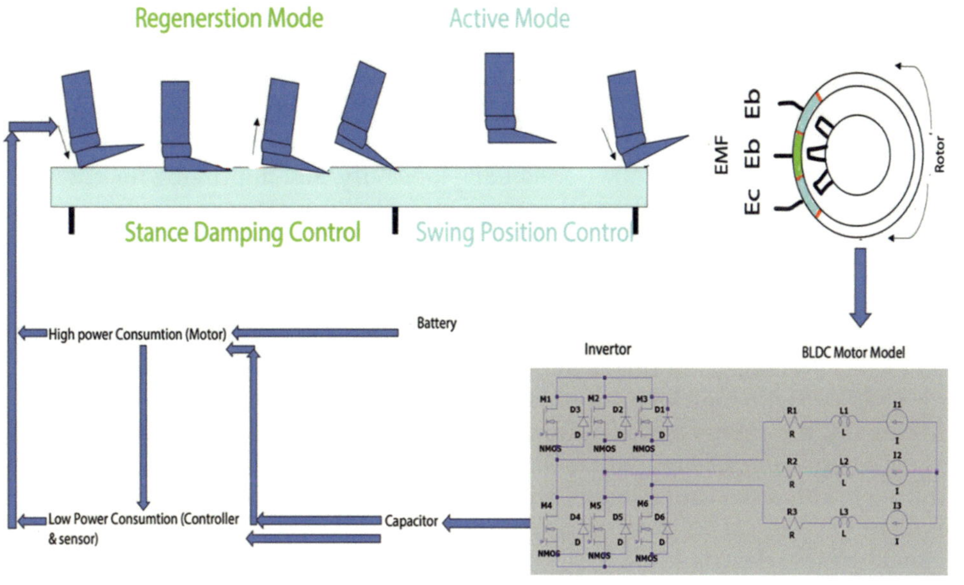

Figure 5.
The gait pattern for an electromagnetic power recovery system in a prosthesis foot [49].

the human ankle performs negative work at normal walking speed to support body mass and store elastic energy [50]. Specifically, during the double-support phase, the leading limb absorbs energy, known as collision work, to prevent shock, while the trailing limb generates positive power [51].

The concept of harnessing this negative power through mechanical absorbers during a specific gait phase and using it in other phases to enhance walking energy efficiency has been explored. Numerous studies have investigated the use of springs to store energy in the early and middle stance phases and release it to propel amputees during the late stance phase [44]. Parallel and series elasticity springs have been incorporated into a robotic prosthesis to capture mechanical energy, reducing the average metabolic cost for amputees by 14% at a self-selected speed. Furthermore, a "plantar flexion" spring and "push-off spring" were employed to store energy and supply additional energy to propel forward walking amputees, such as the Ankle Mimic Prosthetic Foot 2.0 [52]. By combining the energy-storing spring and clutches, the artificial foot could reduce the push-off force and contribute to the reduction of metabolic cost associated with ankle impairments through energy recycling.

4.1.2 Active energy storage systems

The research mentioned has effectively utilized additional mechanical designs to store energy and propel human push-off movements. However, wearable robotic devices often require an electrical energy supply, which adds complexity and potential failure points compared to purely mechanical systems. This raises the question of whether the collected energy can be directly used by these electrical devices without the need for additional mechanical structures. Previous attempts have utilized energy-storage insoles, categorizing the principle of energy storage into two types: electromagnetic [43] and piezoelectric [53].

4.1.2.1 Piezoelectric energy harvesting

Piezoelectricity is the property of materials that allows the conversion of mechanical vibrations into electrical energy. This technology can be incorporated into robotic prostheses to power mechanical limbs, offering several benefits such as reducing the dependence on external batteries, enhancing longevity, and improving functionality compared to conventional prostheses reliant on batteries that require frequent replacement.

Piezoelectric shoe harvesters provide a perpetual power supply, minimizing the need for battery refills and enhancing user convenience [54]. Despite the continued use of external batteries, piezoelectric shoe inserts can substantially prolong their lifetime by harnessing additional energy. This may be particularly advantageous for individuals with high levels of physical activity or who travel frequently [55]. Additionally, the energy obtained from piezoelectric materials can be used to operate various components of the artificial limb, including sensors, motors, and actuators, resulting in enhanced functionality and a more authentic walking experience for the user [34].

Smaller piezoelectric stacks are preferred for their ability to amplify stress inside the stack, resulting in greater power output. Power storage devices and management circuits are essential for practical applications [56]. Treadmill tests were conducted as an initial demonstration of the concept, using a prototype equipped with eight and six piezoelectric stacks. The results showed that the harvester with a lower number of stacks can produce a greater amount of power per stack due to a larger force input, suggesting a trade-off between the number of stacks and the power output [57]. Possible enhancements for the future involve integrating a two-stage force amplification structure and using advanced piezoelectric materials [53].

Substituting flat piezoelectric plates/disks with arc-shaped piezoelectric structures can greatly improve the efficiency of energy harvesters. The findings of this study [58] indicate that arc-shaped piezoelectric energy harvesters (PEHs) exhibit low resonance frequencies, experience uniform stress distribution, and produce higher output voltage, power, power density, and energy conversion efficiency. However, there are still drawbacks in terms of durability and efficiency. The piezoelectric materials used for a robotic prosthesis must possess the capacity to endure the substantial pressures and tensions exerted while walking and running [59]. The harvester must provide sufficient energy to operate the specific functionalities of the prosthesis limb [60].

4.1.2.2 Electromagnetic energy recycling

Electromagnetic energy recycling in robotic prostheses aims to improve energy efficiency by effectively using and retrieving electromagnetic energy within the system [43, 46]. This approach addresses the challenges related to power consumption in robotic prosthetic devices, with the goal of extending their operational periods and enhancing user experience.

The process involves converting lost energy from braking or deceleration into electricity, which is then stored or diverted for future use. Regenerative braking devices play a key role in this technique, as they recover kinetic energy during deceleration and convert it into electricity. This can significantly reduce fuel consumption and extend the life of brake pads [61].

To achieve this, low carbonization, intelligence, and information technology are replacing passive energy usage in vehicle suspension systems with energy harvesting. The development of electromagnetic suspensions for energy regeneration involves complex considerations, such as lightweight design, motor dead zones, and energy regeneration. Future advancements in this field require improvements in structural dependability, energy efficiency, and control [62].

The electromagnetic energy regenerative suspension system is characterized by multifunctionality and parameter uncertainty. Optimization design of such systems involves the use of a multidisciplinary coupling model to analyze system performance. Uncertainty analysis helps clarify the correlation of performance parameters, and integrating optimization techniques can enhance design efficiency. To create a more realistic system, dynamic modeling and improved algorithm stability are needed to reduce local convergence and optimization distortion [62].

Table 1 illustrates a shoe-mounted electromagnetic energy harvesting system [43]. Wave springs are utilized to effectively convert kinetic energy from walking and running into electrical energy. This approach addresses issues faced by current footstep energy harvesting systems. Wave springs, which function similar to coil springs but are smaller, are employed for this purpose. The harvester's top plate is supported by two wave springs that activate when moved, converting the force into rotation using a rack-and-pinion mechanism. Wave spring selection is determined by load capacity and dimensions.

4.2 Actuation system

The actuator component of a hand prosthesis serves as the power source for the device. The selection of an appropriate actuation system has a significant impact on the grasping force and reaction time [63]. Hand prostheses typically utilize one of the two types of power sources: body-powered and externally powered systems.

4.2.1 Body-powered

Body-powered actuation systems require the amputee to generate the grasping force themselves, often incorporating hydraulic cylinders to assist in producing the required force [25]. However, body-powered prosthesis hands are limited to a few grasping patterns due to the restricted power inputs from the user. The real world requires a significant number of varied and complex gripping styles, as seen in **Figures 6** and 7.

4.2.2 Externally powered

In recent times, there has been a shift toward externally powered prosthesis hands to overcome limitations in traditional designs. These innovative actuation systems are

Feature	Description
Technology	Electromagnetic energy harvester
Power source	Kinetic energy from footsteps
Energy output	55.6 μJ per step
Power output	1.6 mW (peak)

Table 1.
Energy output of an electromagnetic energy recovery system in a prosthesis foot.

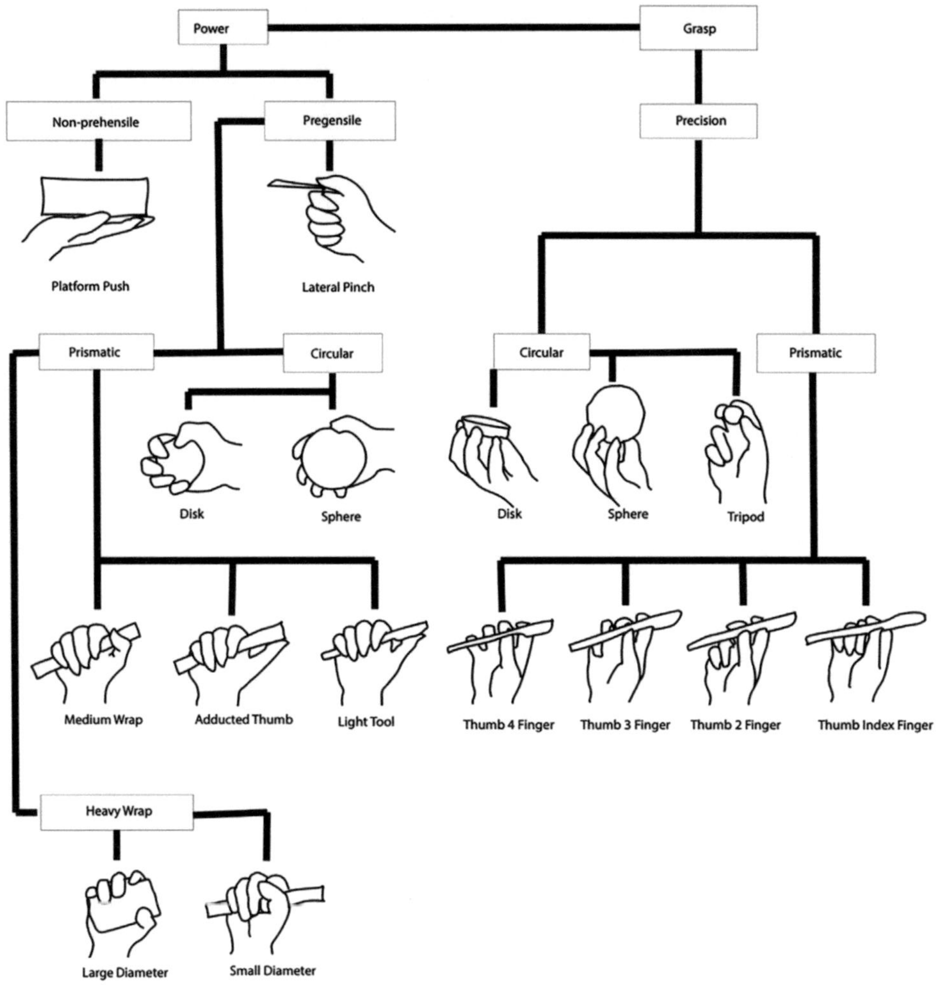

Figure 6.
Possible grasping motions possible with the human hand.

Figure 7.
Bone passions of two different methods of grasping with two fingers.

developed to provide superior grasping functionality and can be classified into three categories: (a) electric motors, (b) artificial muscles, and (c) pneumatic and fluid actuation systems. These systems offer a promising avenue for enhancing the functionality and versatility of prosthesis hands [26].

Initially, myoelectric prosthesis hands were designed for single-degree-of-freedom (DOF) grasping gestures, primarily involving thumb adduction against the other four digits. However, technological advancements have led to the integration of multiple actuators into these prostheses to simulate multigrasping patterns, resembling the functionality of a human hand. Despite these improvements, the addition of multiple actuators has presented challenges related to the device's cost, weight, size, and maintenance. Consequently, many research groups have shifted their focus toward innovating power transmission mechanisms, rather than the power source itself. Differential power transmission mechanisms and underactuated digits have been extensively researched and developed over the past few decades.

4.2.2.1 Electric motors

Electric motors have long been the actuator of choice in prostheses due to their affordability, accessibility, and well-developed nature. They offer a wide range of torque, speed, and power. Direct current (DC) motors, available in brushed and brushless variants, are commonly used. Brushless DC motors (BLDC), including stepper motors, are preferred for their high speed, torque, and efficiency. Stepper motors, a type of BLDC, provide precise motion control and are used in prostheses to enable accurate grasping gestures [64].

Despite their popularity, electric motors require a high reduction ratio gear train to provide sufficient torque for grasping. This increases the size of the motors and creates challenges in fitting them within prostheses. Micro gear DC motors, which can fit inside prosthesis hands without increasing the device's size or cost, are commonly used actuators. These motors are typically installed in the prosthesis's arm, palm, or fingers.

Ultrasonic motors, known for their high output power relative to their small size, have also been used as actuators. Prosthesis hands may incorporate varying numbers of actuators. For example, the Tact hand [65] uses separate actuators to operate each finger individually, offering more flexibility in motion control. However, as the number of DC motors increases, the size of each motor must decrease, resulting in less grasping force. To address this issue, underactuated and differential mechanisms have been developed.

4.2.2.2 Artificial muscles

Electric motors are effective for simple grasping functions in prostheses but become less practical for complex grasping gestures due to the need for multiple motors, increasing the device's weight and reducing its functionality. To address this, researchers have developed soft actuators, or artificial muscles, inspired by the biological function of muscles.

These artificial muscles offer the advantage of providing multiple flexible actuators without increasing the size and weight of the device, operating silently compared to electric motors. Various materials, such as piezoelectric, mechanochemical polymers, conducting polymers, and electromagnetic materials, have been used to create artificial muscles for robotic applications. However, many of these materials are not functional due to their low output power [66].

Shape memory alloys (SMAs) are another potential material for artificial muscles, offering a larger force-to-mass ratio compared to polymer-type actuators [67]. Lee et al. used multiple SMA artificial muscles and increased their length using designated paths to achieve sufficient output force and range of motion. Similarly, She et al. designed a soft robotic finger using SMA actuators.

Yoter et al. developed a hydraulically amplified self-healing electrostatic (HASEL) actuator as an artificial muscle for prosthesis hands, demonstrating that the HASEL actuator is 10.6 times faster and consumes less electrical energy than DC motor actuators [68].

Moreover, fluid systems can be used to create artificial muscles. For example, Kargov et al. [69] developed an artificial muscle made of twisted and coiled nylon fibers. The actuation system circulated hot and cold water to power the prosthesis hand, resulting in rapid finger movement.

4.2.2.3 Pneumatics and hydraulic systems

Rigid structure hand prostheses require intricate mechanisms to perform complex grasp synergies [31]. Some researchers have aimed to increase the prosthesis's degree of freedom (DOF) to offer more flexibility [70]. However, this approach of adding sensors and mechanisms to compensate for the lack of mechanical compliance can lead to a sophisticated and costly design.

The emergence of soft robots has provided a solution, offering mechanical compliance that allows adaptation to the shape of unknown environments without the need for additional sensors and complex mechanisms. Soft materials have been used to create flexible structures for hand prostheses, necessitating flexible actuators.

Current developments in pneumatic and hydraulic actuators are providing soft structures as actuators. Hydraulic systems can be employed in prostheses with multiple DOFs to address the need for numerous electric actuators and bulky prostheses. For instance, Fras and Althoefer [71] developed a soft biomimetic prosthesis hand using a pneumatic system to independently actuate each finger, resulting in a human-like hand prosthesis capable of passively adapting its shape to the object it was grasping (**Figure 8**).

They used conical three-dimensional (3D)-printed cores, wound and coated with thin silicone layers using a mold, and tightly connected to the pneumatic inlet pipes. Similarly, Kargov et al. [69] introduced a miniaturized hydraulic system used as an actuation system for artificial hands, using biocompatible silicone oil as a fluid and embedding micropumps and a fluid reservoir inside the prosthesis's palm. This study demonstrated the device's adaptability during the grasping and holding of objects compared to conventional electromechanical actuation systems.

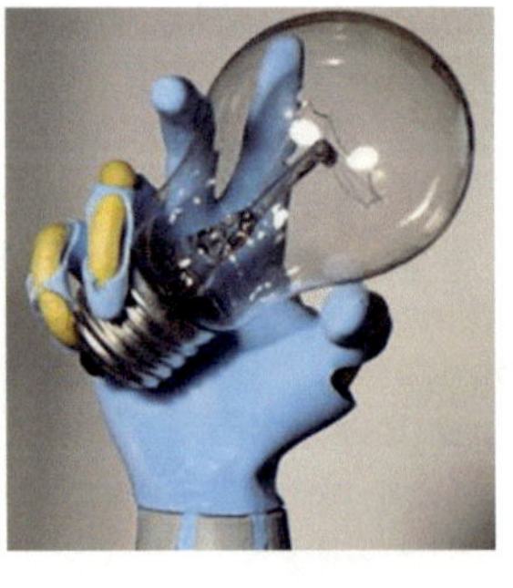

Figure 8.
Soft prosthesis hand [74].

Hybrid actuators that combine the benefits of both electric motors and artificial muscles have also been introduced. For instance, Gao et al. designed a hybrid actuator combining shape memory alloy (SMA) and micro-DC motors, where the micro-DC motors provided sufficient grasping force, and the SMA actuators improved the reflex speed to prevent grasped objects from slipping.

4.3 Mechanisms

Different mechanisms have been used to produce prosthesis devices over the years to increase force, accommodate various activities, and enhance the range of grasping gestures. The mechanism can be divided into three, which are linkage-based mechanisms, cable and pulley mechanisms, and rack-and-pinion mechanisms [72–76].

4.3.1 Linkage-based

Linkage-based mechanisms, such as four-bar linkage prosthesis knee mechanisms, are commonly used in prosthetics to simulate the normal biological knee function for individuals with transfemoral amputation. While these mechanisms offer functional advantages to certain amputees, they are fitted in a limited number of cases [77].

In the context of Knee-Ankle-Foot Orthoses (KAFOs), a study evaluated the efficacy of two different medial linkage mechanisms: a sliding mechanism (SM) and a medial linkage mechanism associated with reciprocating gait motion (MLRGM). The study found that a KAFO fitted with the MLRGM resulted in improvements in walking speed, distance walked, and energy costs compared to the more standard SM. However, the time required for KAFO increased with the use of the MLRGM, while doffing time was approximately equal for both types [72]. These findings suggest that the use of a KAFO with an MLRGM could provide significant benefits for patients with spinal cord injury compared to a KAFO with a standard SM. However, the choice of mechanism should be based on the individual needs and capabilities of the patient [72, 77, 78].

Yao et al. [79] developed a prosthesis hand capable of performing neutral, power, and precision gestures using just one actuator. This design employed a linkage-based mechanism to provide flexion motions for precision grip and dwell action for power grip for the thumb, while the motion of other digits remained identical in each pattern. The study demonstrated a significant increase in grasping force due to the use of a powerful actuator.

4.3.2 Cable-pulley

Cable and pulley mechanisms are commonly used in prosthesis devices, employing a cable and pulley to transmit power, which is particularly advantageous in surgical robots. However, these mechanisms are subject to backlash hysteresis, which presents a challenge in precise position control during surgical procedures. A study by Liang et al. [76] proposed a position compensation scheme using a support vector machine based on feedforward control to reduce position tracking error. The study demonstrated that the scheme significantly improved the positioning precision of the robotic end-effector.

In another study, Catalano et al. [80] introduced a novel differential mechanism capable of providing two degrees of freedom (DOFs) from a single actuator. Their designed prosthesis hand could perform three types of grasping (lateral, precision,

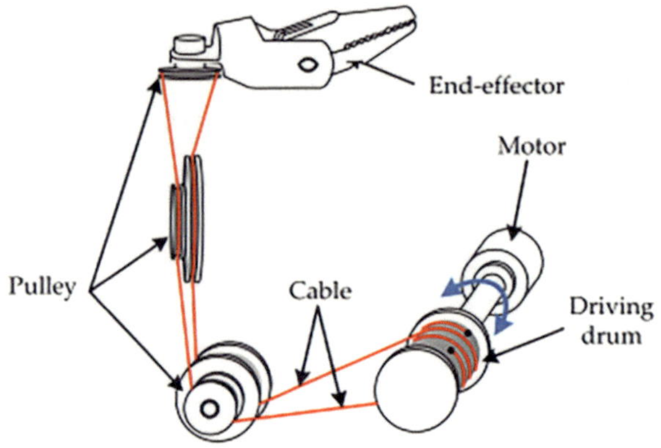

Figure 9.
Cable and pulley used in laparoscopic surgical robot [76].

and power) with a single actuator. The differential mechanism utilized a cable and pulley system to alter the motion of the thumb, enabling different grasp gestures. While the designed prosthesis was lighter in weight, there was a latency due to the switching mechanism (**Figure 9**).

4.3.3 Rack and pinion

Rack-and-pinion mechanisms find application in various fields, including prosthetics, where they convert rotational motion into linear motion, offering advantages for prosthetic devices. In a study by Quinlan et al. [74], a rack-and-pinion design was employed to assess the texturing of prosthesis sockets, aiming to enhance suspension in the transverse plane and reduce rotation. The research revealed that certain texture patterns notably increased torque, providing resistance against unwanted rotation in the transverse plane, particularly for passive suction sockets.

In another study, Xu et al. developed a single-actuator hand prosthesis utilizing a continuum differential mechanism. This design integrated a rack-and-pinion system into a mechanical linkage system to enable continuum motion for the thumb. The prototype prosthesis demonstrated successful grasping of objects with varying shapes (**Figure 10**) [81–83].

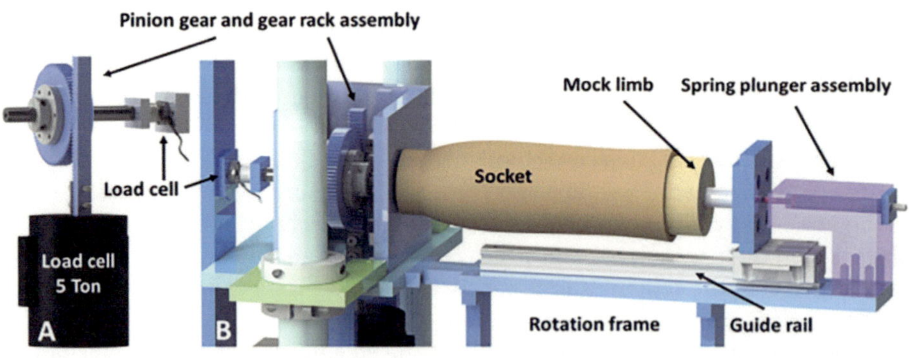

Figure 10.
Rack-and-pinion mechanism [74].

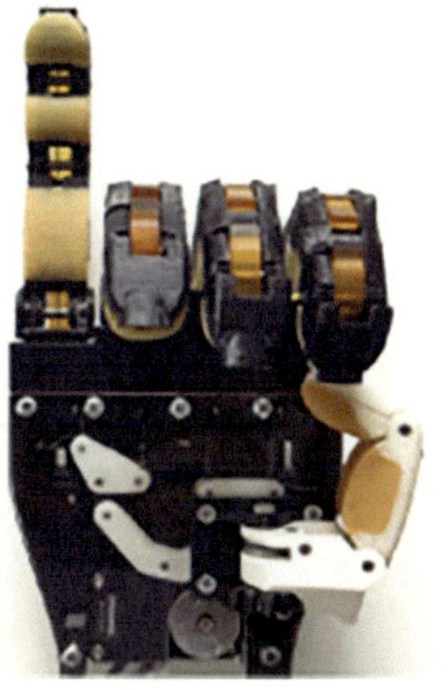

Figure 11.
Prosthesis hand gestures equipped with lockable mechanism [84].

4.3.4 Lockable

Gao et al. [84] designed a selectively lockable differential mechanism to control the angular position of each finger, enabling a wide range of grasping gestures. Their study developed a belt cam differential mechanism for all fingers except the thumb, providing two different positions. Additionally, 6 lockable positions were intended for the thumb, allowing for 96 position combinations in total. This selectively lockable differential mechanism enables the hand to execute various grasping postures, optimizing grasping performance for different objects (**Figure 11**).

4.4 Materials

The evolution of materials used in the construction of upper extremity prosthetic limbs has been significant over time. Initially, wood was the primary material used for replacing lost limbs. However, with technological advancements, there has been a substantial improvement in the materials used for prosthetic limbs. Biomaterials, engineered materials designed for applications in biology and medicine, have played a crucial role in this evolution. When it comes to prostheses, the materials used must be biocompatible to ensure they do not cause harm to living tissue (as shown in **Figure 12**).

The selection of suitable materials for prosthetic use involves considering a broad range of mechanical, physical, and material properties, including wear resistance, yield strength, and ductility. This careful selection process is crucial to ensure the safety, durability, and functionality of the prosthetic limb [38].

4.4.1 Metal

Titanium and aluminum are the most commonly used metals in the skeletal design of upper extremity prostheses, although other metals such as copper, iron, magnesium, and steel are also used to some extent. Titanium is particularly favored for its advantageous properties that make it suitable for biomedical applications, including excellent corrosion resistance, light weight, and a favorable strength-to-weight ratio [37].

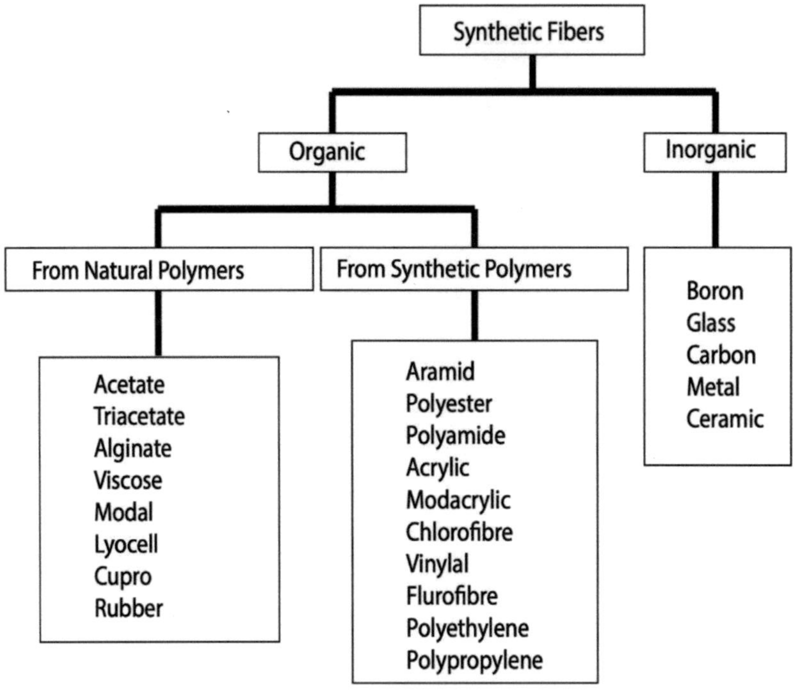

Figure 12.
Types and categories of materials used in the production of prosthesis limbs.

To enhance its mechanical properties, titanium is often alloyed with other metals such as aluminum and vanadium. Furthermore, the low modulus of elasticity of titanium is similar to that of human bone, allowing for a more uniform distribution of the user's skeletal load between the implant and the bone, resulting in improved gait for the user.

Despite the many benefits of titanium as a biomaterial, its high cost can be a barrier to access, particularly in countries without a comprehensive healthcare system. This highlights the need for ongoing research and development to find cost-effective alternatives without compromising the quality and performance of prosthetic limbs.

4.4.2 Plastics and polymers

The socket, a critical component of a prosthesis, is meticulously crafted to fit the user's residual limb. It's created based on a precise mold of the remaining limb, ensuring an accurate fit. This process is so crucial that it necessitates the expertise of an experienced prosthetist.

The socket's structure is typically made from thermoplastics, including polyester resin or acrylic resin. Other polymers, such as nylon and acrylics, are also used, particularly for prosthesis socks that cover the prosthesis. Furthermore, polyurethane foam is widely used in various parts of the prosthesis, from soft, cosmetic foam covers to rigid structural sections.

This information provides a glimpse into the complex and intricate world of prostheses, highlighting the importance of materials science and skilled craftsmanship in creating functional, comfortable prosthesis devices [5, 85].

Additive manufacturing, also known as 3D printing, has gained popularity due to its affordability and the convenience it offers in rapid prototyping. It allows for the swift production of custom-designed parts using cost-effective thermoplastics and resins. Fused deposition modeling (FDM), a prevalent method in additive manufacturing, employs an extruder head to deposit filament layer by layer, resulting in the desired part. Acrylonitrile butadiene styrene (ABS), a lightweight thermoplastic known for its durability and toughness, is frequently used in this process. This has led to the widespread use of 3D-printed ABS in open-source prosthesis hands such as the Cyborg-Beast or the Raptor Hand [86, 87]. These products aim to provide a low-cost, customizable prosthesis solution. However, despite the accessibility of custom prosthesis designs through 3D printing, these devices often lack the strength and durability required for practical use, resulting in several disadvantages when compared to commercially manufactured prosthesis devices.

4.4.3 Carbon composites

Carbon fiber is a material that's highly sought after in the design of prosthesis pylons, and for good reason. Its unique properties make it an ideal choice for such applications. Carbon fiber is renowned for its high strength and lightweight, making it perfect for creating durable yet comfortable prostheses. Moreover, carbon fiber exhibits high chemical and temperature resistance, which means it can withstand various environmental conditions without degrading. Its high stiffness and low thermal expansion further enhance its suitability for prosthesis applications, as these properties ensure the stability and reliability of the prosthesis. The specific modulus of carbon fiber is approximately three times higher than those of magnesium and titanium. This means that carbon fiber can bear more stress and strain before deforming, making it a superior material in terms of mechanical performance. However, it's worth noting that the cost of carbon fiber can be high when compared to other materials with similar mechanical properties. This is due to the complex manufacturing process, which involves the oxidation, carbonization, and graphitization of the carbon precursor polyacrylonitrile [2, 37].

5. Conclusions

This paper provides a review of prosthetic robots from various perspectives, focusing on the state-of-the-art and the challenges in gaining an overall understanding of prosthetic development from 1980s to the present day. It demonstrates that current research is moving toward creating multifunctional and efficient prosthetic systems. Existing research has improved control systems, sensing, actuation, and mechanisms, all of which contribute to the overall prosthetic system's effectiveness. There is great potential for expansion as technology continues to develop and the demand for prosthetics in real human life increases. Artificial intelligence and high-end sensor technology should be integrated into the current system. In conclusion, the development of prosthetic robots has shown steady progress, but there is still much room for improvement. By studying and understanding the structure and capabilities of human limbs, researchers and engineers can further advance robotic prosthetic technology, aiming for even greater functionality and robustness to assist humans.

The following are recommendations for future researchers to explore in expanding this area:

5.1 Recommendations

1. Research into power-efficient prostheses should continue to enhance mobility and independence by assisting individuals with amputations in performing daily activities, from simple to complex tasks [88, 89]. This, in turn, improves their quality of life [26, 79, 90]. Additionally, power efficiency in robotic prostheses contributes to the longevity and durability of the device, reducing the need for frequent replacements or repairs. This lowers the overall physical and psychological burden on the user [5, 37–38, 85]. Energy consumption can also be reduced, considering the limited battery capacity of these devices [40, 41].

   Therefore, the development of sophisticated energy management systems is essential to ensure optimal utilization of stored energy, prolonging the operational duration of the prosthesis without compromising its performance. These systems involve intelligent control algorithms and feedback mechanisms to adapt to the user's movements and optimize power consumption in real time [2, 29, 30]. As the field of robotic prostheses continues to evolve, interdisciplinary approaches that leverage expertise in materials science, robotics, and bioengineering will be instrumental in realizing the next generation of power-efficient and seamlessly integrated robotic prosthesis devices [1, 25, 26, 31].

2. As research in the field of robotic prostheses continues to advance, ongoing efforts in material science, biomechanics, and design optimization will play a pivotal role in addressing the weight balance challenge. Interdisciplinary collaborations and a comprehensive understanding of human physiology and biomechanics will drive the development of next-generation robotic prostheses that prioritize both power efficiency and weight balance, ultimately improving the quality of life for individuals in need of these advanced mobility solutions [2, 36].

   Researchers and manufacturers aim to strike a balance between these factors to enhance the user experience and functionality of the prosthesis devices. They are continually innovating to find materials and designs that can optimize power efficiency, enhance mobility, and improve the longevity and durability of these devices.

3. The use of energy regeneration systems can increase the longevity of prostheses. Research has shown up to a 30% increase in the battery life of the limb through the use of these systems. Energy efficiency can be further improved by embedding an electromagnetic energy or piezoelectric approach in the system [40–42].

4. The choice of hybrid actuators can be considered effective and practical, as they can reduce weight while maintaining functionality. For example, a combination of an electrical motor and artificial muscles can be used in the complex design of prostheses. The electrical motor is highly resistant to environmental factors and is lighter than other alternatives.

5. The material chosen for the structure of prostheses must be suitable to give the user a feeling and environment similar to a real limb. Carbon fiber is commonly used for the external segment of prostheses, while titanium is used for bone attachment. This choice allows for lighter, more compact, permanent, and chemically and environmentally resistant prostheses, but it requires invasive surgeries for attachment.

   Additionally, 3D printing and additive manufacturing technology are revolutionizing the field of prostheses by enabling the creation of custom-fit devices. This technology can significantly reduce costs and increase accessibility to prostheses for those in lower-income and less-developed countries. It allows amputees to customize the fit and feel of their prostheses and enables rapid iteration at low cost.

   Ongoing research is focusing on the development of advanced coating and surface modification techniques, such as the use of bioactive coatings like titanium and carbon fiber. These coatings can promote the integration of the prosthesis with surrounding tissues, improve its biocompatibility, and reduce the body's immune response and degradation of the prosthesis within the body [91].

6. The integration of artificial intelligence (AI) and machine learning (ML) into prostheses can significantly improve their functionality, making them smarter, more responsive, and easier to control. These technologies can learn over time, adapting to the user's routines and habits to become more efficient and attuned to their needs. Continued research in this area is expected to lead to even more innovative solutions that will enhance the quality of life for individuals who use prosthetic devices.

   Artificial intelligence serves as the foundation of robotic prosthetic limbs. These AI systems, comprising interconnected parallel processing elements, work together to address various challenges [88]. When used in rehabilitation devices, they offer significant advantages, such as adaptive learning, self-organization, real-time operation, fault tolerance through redundant information coding, and the ability to learn relationships between variables [89].

7. Another area of research is the development of prosthetic limbs that can be consciously controlled by the user. These devices not only allow for more natural movement but also provide the user with a sense of touch by connecting directly to the remaining nerves in the amputated limb [92].

8. Enhancing the clinical use of robotic prosthetic limbs involves overcoming several hurdles, including the cost of research and commercialization, the complexity of signal acquisition systems, the high expense of computational methods, and the need for robust devices. The primary strategy is to reduce the cost of current solutions while simultaneously increasing the robustness of the prosthesis. This can be achieved through the use of stronger materials, more powerful and versatile control systems, and longer use times enabled by regenerative systems.

Acknowledgements

This work was supported by Rabdan Academy under grant no (to be confirmed later).

Conflict of interest

The authors declare no conflict of interest.

Author details

Nohaidda Sariff[1,2]*, Denesh Sooriamoorthy[3], Ahmad Shah Hizam Md Yasir[4], Puteri Nor Aznie Fahsyar Syed Mahadzir[1], Joy Massouh[3], Miqdad Taqi Mohamed Mushadiq[3], Julian Tan Kok Ping[1] and Steven Eu Kok Seng[1]

1 Faculty of Innovation and Technology, Taylor's University, Subang Jaya, Selangor, Malaysia

2 Digital Health and Medical Advancement Impact Lab, Taylor's University, Subang Jaya, Selangor, Malaysia

3 School of Engineering, Asia Pacific University, Kuala Lumpur, Malaysia

4 Faculty of Resilience, Rabdan Academy, Abu Dhabi, United Arab Emirates

*Address all correspondence to: nohaidda.sariff@taylors.edu.my

References

[1] Laschowski B, McPhee J. Energy-efficient actuator design principles for robotic leg prostheses and exoskeletons: A review of series elasticity and backdrivability. Journal of Computational and Nonlinear Dynamics. 2023;**18**(6):060801, pp. 1-8. DOI: 10.1115/1.4056919

[2] Gabert L, Hood S, Tran M, Cempini M, Lenzi T. A compact, lightweight robotic ankle-foot prosthesis: Featuring a powered polycentric design. IEEE Robotics and Automation Magazine. 2020;**27**(1):87-102. DOI: 10.1109/MRA.2019.2955740

[3] Nerlich AG, Zink A, Szeimies U, Hagedorn HG. Ancient Egyptian prosthesis of the big toe. The Lancet. 2000;**356**(9248):2176-2179. DOI: 10.1016/S0140-6736(00)03507-8

[4] Laschowski B. Energy Regeneration and Environment Sensing for Robotic Leg Prostheses and Exoskeletons. [thesis] 2021

[5] Brack R, Amalu EH. A review of technology, materials and R&D challenges of upper limb prosthesis for improved user suitability. Journal of Orthopaedics. 25 Dec 2020;**23**:88-96. DOI: 10.1016/j.jor.2020.12.009. PMID: 33442223, PMCID: PMC7787923

[6] Mendez V, Iberite F, Shokur S, Micera S. Current solutions and future trends for robotic prosthetic hands. Annual Review of Control, Robotics, and Autonomous Systems. 2021;**4**(1):595-627. DOI: 10.1146/annurev-control-071020-104336

[7] Price MA, Beckerle P, Sup FC. Design optimization in lower limb prostheses: A review. IEEE Transactions on Neural Systems and Rehabilitation Engineering. 2019;**27**(8):1574-1588. DOI: 10.1109/TNSRE.2019.2927094

[8] Babu D, Nasir A, Jamaludin AS, Rosle MH. Holding, Grasping and Sensing of Prosthetic Robot Arm Like a Real Human Hand, a Journey Beyond Limits: An Extensive Review. 2022. pp. 485-504. DOI: 10.1007/978-981-16-4115-2_39

[9] Sariff N, Hilmi Ismail Z, Yasir ASHM, Sooriamoorthy D, Mahadzir PNAFS. Multi-agent robot motion planning for rendezvous applications in a mixed environment with a broadcast event-triggered consensus controller. In: Motion Planning for Dynamic Agents [Working Title]. London, UK: IntechOpen; 2023. DOI: 10.5772/intechopen.1002494

[10] Baggetta M, Palli G, Melchiorri C, Berselli G. A monolithic cable-driven compliant wrist for prosthetic arms. IEEE/ASME Transactions on Mechatronics. 2024;**29**(2):10385249.1052-10385249.1063. DOI: 10.1109/TMECH.2023.3345215

[11] Wang H, Tao Q, Zhang XD. Ensemble learning method for the continuous decoding of hand joint angles. Sensors. 2024;**24**(2):1-17. DOI: 10.3390/s24020660

[12] Toedtheide A, Fortunic EP, Kühn J, Jensen E, Haddadin S. A transhumeral prosthesis with an artificial neuromuscular system: Sim2real-guided design, modeling, and control. International Journal of Robotics Research. 2024;**43**(7):942-980. DOI: 10.1177/02783649231218719

[13] Mendez J, Murray R, Gabert L, Fey NP, Liu HH, Lenzi T. Continuous

A-mode ultrasound-based prediction of transfemoral amputee prosthesis kinematics across different ambulation tasks. IEEE Transactions on Biomedical Engineering. 2024;**71**(1):56-67. DOI: 10.1109/TBME.2023.3292032

[14] Mazumder A, Hekman EEG, Carloni R. Toward controlling transtibial prostheses using a single degree of freedom inertial sensor system. IEEE Access. 2024;**12**:24803-24812. DOI: 10.1109/ACCESS.2024.3364498

[15] Yuan MC, Ling TX, Su Q, Wan XF, Lai YH, Zhou ZK. Safety and effectiveness of robotic-arm assisted total knee arthroplasty. Orthopaedic Surgery. 2024;**16**(4):882-893. DOI: 10.1111/os.14008

[16] Juan MC, Hidaldo C, Mifsut D. A mixed reality application for total hip arthroplasty. Virtual Reality. 2024;**28**(1):1-14. DOI: 10.1007/s10055-024-00938-9

[17] Buchan GBJ, Hecht CJII, Rodriguez-Elizalde S, Kabata T, Kamath AF. Automated digital templating of component sizing is accurate in robotic total hip arthroplasty when compared to predicate software. Medical Engineering & Physics. 2024;**124**:1-5. DOI: 10.1016/j.medengphy.2024.104105

[18] Cifuentes-Cuadros AA, Romero E, Caballa S, Vega-Centeno D, Elias DA. The LIBRA NeuroLimb: Hybrid real-time control and mechatronic design for affordable prosthetics in developing regions. Sensors. 2024;**24**(1):1-37. DOI: 10.3390/s24010070

[19] Ploumitsakou M et al. Remapping wetness perception in upper limb amputees. Advanced Intelligent Systems. 2024;**6**(3):2300512. DOI: 10.1002/aisy.202300512

[20] Gonzalez MA et al. Electrical stimulation of regenerative peripheral nerve interfaces (RPNIs) induces referred sensations in people with upper limb loss. IEEE Transactions on Neural Systems and Rehabilitation Engineering. 2024;**32**:339-349. DOI: 10.1109/TNSRE.2023.3345164

[21] Capsi-Morales P, Piazza C, Grioli G, Bicchi A, Catalano MG. A three degrees of freedom switchable impedance myoelectric prosthetic wrist. The International Journal of Robotics Research. 2024. DOI: 10.1177/02783649241231298

[22] Lee C, Vaskov AK, Davis AJ, Kartes JM, Gates DH. Development and validation of the coffee task: A novel functional assessment for prosthetic grip selection. Journal of Neuroengineering and Rehabilitation. 2024;**21**(1):1-10. DOI: 10.1186/s12984-024-01307-y

[23] Han MS, Harnett CK. Journey from human hands to robot hands: Biological inspiration of anthropomorphic robotic manipulators. Bioinspiration & Biomimetics. 2024;**19**(2):1-15. DOI: 10.1088/1748-3190/ad262c

[24] Herrin KR, Kwak ST, Rock CG, Chang YH. Gait quality in prosthesis users is reflected by force-based metrics when learning to walk on a new research-grade powered prosthesis. Frontiers in Rehabilitation Sciences. 2024;**5**:1-11. DOI: 10.3389/fresc.2024.1339856

[25] Smit G, Plettenburg DH, van der Helm FCT. The lightweight delft cylinder hand: First multi-articulating hand that meets the basic user requirements. IEEE Transactions on Neural Systems and Rehabilitation Engineering. 2015;**23**(3):431-440. DOI: 10.1109/TNSRE.2014.2342158

[26] Moeinnia H, Su H, Kim WS. Novel grasping mechanisms of 3D-printed prosthetic hands. Advanced Intelligent Systems. 2022;**4**(11):1-16. DOI: 10.1002/aisy.202200189

[27] Wu L, Xie S, Zhang Z, Meng W. Energy efficiency of gait rehabilitation robot: A review. In: 2019 IEEE/ASME International Conference on Advanced Intelligent Mechatronics (AIM). IEEE; 2019. pp. 465-470. DOI: 10.1109/AIM.2019.8868444

[28] Verstraten T, Geeroms J, Mathijssen G, Convens B, Vanderborght B, Lefeber D. Optimizing the power and energy consumption of powered prosthetic ankles with series and parallel elasticity. Mechanism and Machine Theory. 2017;**116**:419-432. DOI: 10.1016/j.mechmachtheory.2017.06.004

[29] Kashiri N et al. An overview on principles for energy efficient robot locomotion. Frontiers in Robotics and AI. 2018;**5**:1-13. DOI: 10.3389/frobt.2018.00129

[30] Hitt J, Sugar T, Holgate M, Bellman R, Hollander K. Robotic transtibial prosthesis with biomechanical energy regeneration. Industrial Robot: An International Journal. 2009;**36**(5):441-447. DOI: 10.1108/01439910910980169

[31] Gopura RARC, Bandara DSV. A hand prosthesis with an under-actuated and self-adaptive finger mechanism. Engineering. 2018;**10**(07):448-463. DOI: 10.4236/eng.2018.107031

[32] Xie L, Wang Z, Huang G, Liu B, Zhou Z. Mechanical efficiency investigation of an ankle-assisted robot for human walking with a backpack-load. Journal of Biomechanical Engineering. 2021;**143**(11):1-9. DOI: 10.1115/1.4051434

[33] Rohani F, Richter H, van den Bogert AJ. Optimal design and control of an electromechanical transfemoral prosthesis with energy regeneration. PLoS ONE. 2017;**12**(11):e0188266. DOI: 10.1371/journal.pone.0188266

[34] Windrich M, Grimmer M, Christ O, Rinderknecht S, Beckerle P. Active lower limb prosthetics: A systematic review of design issues and solutions. Biomedical Engineering Online. 2016;**15**(S3):140. DOI: 10.1186/s12938-016-0284-9

[35] Tran M, Gabert L, Hood S, Lenzi T. A lightweight robotic leg prosthesis replicating the biomechanics of the knee, ankle, and toe joint. Science robotics. 2022;**7**(72). DOI: 10.1126/scirobotics.abo3996

[36] Ortiz J, Poliero T, Cairoli G, Graf E, Caldwell DG. Energy efficiency analysis and design optimization of an actuation system in a soft modular lower limb exoskeleton. IEEE Robotics and Automation Letters. 2018;**3**(1):484-491. DOI: 10.1109/LRA.2017.2768119

[37] Igual C, Pardo LA Jr, Hahne JM, Igual J. Myoelectric control for upper limb prostheses. Electronics (Basel). 2019;**8**(11):1244. DOI: 10.3390/electronics8111244

[38] dos Santos VTM et al. Characterization of human AB serum for mesenchymal stromal cell expansion. Transfusion Medicine and Hemotherapy. 2017;**44**(1):11-21. DOI: 10.1159/000448196

[39] Roberts TJ. The integrated function of muscles and tendons during locomotion. Comparative Biochemistry and Physiology. Part A, Molecular & Integrative Physiology. 2002;**133**(4):1087-1099. DOI: 10.1016/S1095-6433(02)00244-1

[40] Gates DH. Characterizing Ankle Function during Stair Ascent, Descent, and Level Walking for Ankle Prosthesis and Orthosis Design. Boston University; 2004

[41] Donelan JM, Kram R, Kuo AD. Mechanical work for step-to-step transitions is a major determinant of the metabolic cost of human walking. Journal of Experimental Biology. 2002;**205**(23):3717-3727. DOI: 10.1242/jeb.205.23.3717

[42] Khaligh A, Zeng P, Zheng C. Kinetic energy harvesting using piezoelectric and electromagnetic technologies—State of the art. IEEE Transactions on Industrial Electronics. 2010;**57**(3):850-860. DOI: 10.1109/TIE.2009.2024652

[43] Zhu D, Duarte-Rabelo I, Ayala-Garcia IN, Somov A. An electromagnetic in-shoe energy harvester using wave springs. In: 2018 IEEE Industrial Cyber-Physical Systems (ICPS). Vol. 2018. IEEE. pp. 659-663. DOI: 10.1109/ICPHYS.2018.8390785

[44] Au SK, Weber J, Herr H. Powered ankle—Foot prosthesis improves walking metabolic economy. IEEE Transactions on Robotics. 2009;**25**(1):51-66. DOI: 10.1109/TRO.2008.2008747

[45] Shultz AH, Lawson BE, Goldfarb M. Variable cadence walking and ground adaptive standing with a powered ankle prosthesis. IEEE Transactions on Neural Systems and Rehabilitation Engineering. 2016;**24**(4):495-505. DOI: 10.1109/TNSRE.2015.2428196

[46] Feng Y, Mai J, Agrawal SK, Wang Q. Energy regeneration from electromagnetic induction by human dynamics for lower extremity robotic prostheses. IEEE Transactions on Robotics. 2020;**36**(5):1442-1451. DOI: 10.1109/TRO.2020.2991969

[47] Tucker MR, Fite KB. Mechanical damping with electrical regeneration for a powered transfemoral prosthesis. In: 2010 IEEE/ASME International Conference on Advanced Intelligent Mechatronics. IEEE; 2010. pp. 13-18. DOI: 10.1109/AIM.2010.5695828

[48] Rarick R, Richter H, van den Bogert A, Simon D, Warner H, Barto T. Optimal design of a transfemoral prosthesis with energy storage and regeneration. In: 2014 American Control Conference. Vol. 2014. IEEE. pp. 4108-4113. DOI: 10.1109/ACC.2014.6859051

[49] Wang Q, Yuan K, Zhu J, Wang L. Walk the walk: A lightweight active transtibial prosthesis. IEEE Robotics and Automation Magazine. 2015;**22**(4):80-89. DOI: 10.1109/MRA.2015.2408791

[50] Gates DH, Lelas J, Croce UD, Herr H, Bonato P. Characterization of ankle function during stair ambulation. In: The 26th Annual International Conference of the IEEE Engineering in Medicine and Biology Society. IEEE. pp. 4248-4251. DOI: 10.1109/IEMBS.2004.1404184

[51] Kuo AD, Donelan JM, Ruina A. Energetic consequences of walking like an inverted pendulum: Step-to-step transitions. Exercise and Sport Sciences Reviews. 2005;**33**(2):88-97

[52] Cherelle P, Grosu V, Matthys A, Vanderborght B, Lefeber D. Design and validation of the ankle mimicking prosthetic (AMP-) foot 2.0. IEEE Transactions on Neural Systems and Rehabilitation Engineering. 2014;**22**(1):138-148. DOI: 10.1109/TNSRE.2013.2282416

[53] Qian F, Xu T-B, Zuo L. Design, optimization, modeling and testing of a piezoelectric footwear energy harvester. Energy Conversion and Management.

2018;**171**:1352-1364. DOI: 10.1016/j.enconman.2018.06.069

[54] Anton SR, Sodano HA. A review of power harvesting using piezoelectric materials (2003-2006). Smart Materials and Structures. 2007;**16**(3):R1-R21. DOI: 10.1088/0964-1726/16/3/R01

[55] Qian F, Xu T-B, Zuo L. Piezoelectric energy harvesting from human walking using a two-stage amplification mechanism. Energy. 2019;**189**:116140. DOI: 10.1016/j.energy.2019.116140

[56] Li W, Chen X. Powering body area sensor networks. Matter. 2020;**2**(5):1085-1086. DOI: 10.1016/j.matt.2020.04.006

[57] Fan K, Liu Z, Liu H, Wang L, Zhu Y, Yu B. Scavenging energy from human walking through a shoe-mounted piezoelectric harvester. Applied Physics Letters. 2017;**110**(14). DOI: 10.1063/1.4979832

[58] Yang Z, Wang YQ, Zuo L, Zu J. Introducing arc-shaped piezoelectric elements into energy harvesters. Energy Conversion and Management. 2017;**148**:260-266. DOI: 10.1016/j.enconman.2017.05.073

[59] Rasmussen L. Electrically driven mechanochemical artificial muscle: For smooth three-dimensional movement in robotics and prosthetics. In: proceedings SPIE 6524, Electroactive Polymer Actuators and Devices (EAPAD); 5 Apr 2007:652423. DOI: 10.1117/12.723273

[60] Yang Z, Erturk A, Zu J. On the efficiency of piezoelectric energy harvesters. Extreme Mechanics Letters. 2017;**15**:26-37. DOI: 10.1016/j.eml.2017.05.002

[61] Hamada AT, Orhan MF. An overview of regenerative braking systems. Journal of Energy Storage. 2022;**52**:105033. DOI: 10.1016/j.est.2022.105033

[62] Fu C, Lu J, Ge W, Tan C, Li B. A review of electromagnetic energy regenerative suspension system & key technologies. Computer Modeling in Engineering and Sciences. 2023;**135**(3):1779-1824. DOI: 10.32604/cmes.2022.023092

[63] Rea P. On the design of underactuated finger mechanisms for robotic hands. In: Martinez-Alfaro H, editor. Advances in Mechatronics. Rijeka: IntechOpen; 2011. DOI: 10.5772/24304. (Ch. 6)

[64] González-Vargas J, Ibáñez J, Contreras-Vidal JL, van der Kooij H, Pons JL, editors. Wearable Robotics: Challenges and Trends, Biosystems & Biorobotics. Vol. 16. Cham: Springer International Publishing; 2017. DOI: 10.1007/978-3-319-46532-6

[65] Magrini E, Flacco F, De Luca A. Control of generalized contact motion and force in physical human-robot interaction. In: 2015 IEEE International Conference on Robotics and Automation (ICRA). IEEE; 2015. pp. 2298-2304. DOI: 10.1109/ICRA.2015.7139504

[66] Mirfakhrai T, Madden JDW, Baughman RH. Polymer artificial muscles. Materials Today. 2007;**10**(4):30-38. DOI: 10.1016/S1369-7021(07)70048-2

[67] Liu M, Hao L, Zhang W, Zhao Z. A novel design of shape-memory alloy-based soft robotic gripper with variable stiffness. International Journal of Advanced Robotic Systems. 2020;**17**(1):172988142090781. DOI: 10.1177/1729881420907813

[68] Yoder Z et al. Design of a high-speed prosthetic finger driven by Peano-HASEL

actuators. Frontiers in Robotics and AI. 2020;7:181

[69] Kargov A, Werner T, Pylatiuk C, Schulz S. Development of a miniaturised hydraulic actuation system for artificial hands. Sensors and Actuators, A: Physical. 2008;**141**(2):548-557. DOI: 10.1016/j.sna.2007.10.025

[70] Bandara DSV, Gopura RARC, Hemapala KTMU, Kiguchi K. Development of a multi-DoF transhumeral robotic arm prosthesis. Medical Engineering & Physics. 2017;**48**:131-141. DOI: 10.1016/j.medengphy.2017.06.034

[71] Fras J, Althoefer K. Bio-mimetic pneumatic soft prosthetic hand. Poster Papers. UK-RAS17-29, 2017:90

[72] Seyyedzadeh H, Arazpour M, Saeedi H, Mousavi ME, Golchin N. Comparison of the efficacy of two different medial linkage mechanisms in knee-ankle-foot orthoses on walking ability in subjects with spinal cord injury. JPO Journal of Prosthetics and Orthotics. 2021;**33**(4):311-314. DOI: 10.1097/JPO.0000000000000364

[73] Radcliffe CW. Four-bar linkage prosthetic knee mechanisms. Prosthetics and Orthotics International. 1994;**18**(3):159-173. DOI: 10.3109/03093649409164401

[74] Quinlan J, Subramanian V, Yohay J, Poziembo B, Fatone S. Using mechanical testing to assess texturing of prosthetic sockets to improve suspension in the transverse plane and reduce rotation. PLoS ONE. 2020;**15**(6):e0233148. DOI: 10.1371/journal.pone.0233148

[75] Nuñez NNR, Florez AR, Vieira RS, Martins D. Dimensional synthesis of rack-and-pinion steering mechanism using a novel synthesis equation. Journal of the Brazilian Society of Mechanical Sciences and Engineering. 2023;**45**(8):411. DOI: 10.1007/s40430-023-04317-4

[76] Liang Y, Du Z, Wang W, Sun L. A novel position compensation scheme for cable-pulley mechanisms used in laparoscopic surgical robots. Sensors. 2017;**17**(10):2257. DOI: 10.3390/s17102257

[77] Liang W et al. Mechanisms and component design of prosthetic knees: A review from a biomechanical function perspective. Frontiers in Bioengineering and Biotechnology. 2022;**10**:1-24. DOI: 10.3389/fbioe.2022.950110

[78] Ramirez IA. Linkage-Based Prosthetic Fingertips: Analysis and Testing. Florida Conference on Recent Advances in Robotics, FCRAR 2007. pp. 1-7

[79] Yao S, Ceccarelli M, Zhan Q, Carbone G, Lu Z. Analysis and design of a modular underactuated mechanism for robotic fingers. Proceedings of the Institution of Mechanical Engineers, Part C: Journal of Mechanical Engineering Science. 2012;**226**(1):242-256. DOI: 10.1177/0954406211412457

[80] Catalano MG, Grioli G, Farnioli E, Serio A, Piazza C, Bicchi A. Adaptive synergies for the design and control of the Pisa/IIT SoftHand. International Journal of Robotics Research. 2014;**33**(5):768-782. DOI: 10.1177/0278364913518998

[81] Xu K, Liu H. Continuum differential mechanisms and their applications in gripper designs. IEEE Transactions on Robotics. 2016;**32**(3):754-762. DOI: 10.1109/TRO.2016.2561295

[82] Xu K, Liu H, Zhang Z, Zhu X. Wrist-powered partial hand prosthesis using

a continuum whiffle tree mechanism: A case study. IEEE Transactions on Neural Systems and Rehabilitation Engineering. 2018;**26**(3):609-618. DOI: 10.1109/TNSRE.2018.2800162

[83] Liu H, Zhang Z, Zhu X, Xu K. A single-actuator gripper with a working mode switching mechanism for grasping and rolling manipulation. In: 2018 IEEE/ASME International Conference on Advanced Intelligent Mechatronics (AIM). IEEE; 2018. pp. 359-364. DOI: 10.1109/AIM.2018.8452711

[84] Gao G, Shahmohammadi M, Gerez L, Kontoudis G, Liarokapis M. On differential mechanisms for underactuated, lightweight, adaptive prosthetic hands. Frontiers in Neurorobotics. 2021;**15**:1-12. DOI: 10.3389/fnbot.2021.702031

[85] Cabibihan J-J, Abubasha MK, Thakor N. A method for 3-D printing patient-specific prosthetic arms with high accuracy shape and size. IEEE Access. 2018;**6**:25029-25039. DOI: 10.1109/ACCESS.2018.2825224

[86] Zuniga J et al. Cyborg beast: A low-cost 3d-printed prosthetic hand for children with upper-limb differences. BMC Research Notes. 2015;**8**(1):10. DOI: 10.1186/s13104-015-0971-9

[87] Ma RR, Belter JT, Dollar AM. Hybrid deposition manufacturing: Design strategies for multimaterial mechanisms via three-dimensional printing and material deposition. Journal of Mechanisms and Robotics. 2015;7(2): 1-12. DOI: 10.1115/1.4029400

[88] McGimpsey G, Bradford TC. Limb Prosthetics Services and Devices. Bioengineering Institute Center for Neuroprosthetics Worcester Polytechnic Institution; 2008. pp. 1-35

[89] Jena RK, Aqel MM, Srivastava P, Mahanti PK. Soft computing methodologies in bioinformatics. European Journal of Scientific Research. 2009;**26**(2):189-203

[90] Wu L, Carbone G, Ceccarelli M. Designing an underactuated mechanism for a 1 active DOF finger operation. Mechanism and Machine Theory. 2009;**44**(2):336-348. DOI: 10.1016/j.mechmachtheory.2008.03.011

[91] Kulkarni PG et al. Overcoming challenges and innovations in orthopedic prosthesis design: An interdisciplinary perspective. Biomedical Materials & Devices. 2023;**2**:58-69. DOI: 10.1007/s44174-023-00087-8

[92] Tabot GA et al. Restoring the sense of touch with a prosthetic hand through a brain interface. Proceedings of the National Academy of Sciences. 2013;**110**(45):18279-18284. DOI: 10.1073/pnas.1221113110

Section 2

Robot Kinematics and AI in Robotic Technology

Chapter 4

The Inverse Kinematics of a New Planar Hybrid Robot Manipulator

Serdar Küçük

Abstract

The amount of robots in industrial automation systems is increasing day by day. As a result, the designs of robots are diversifying. The robot industry, which serial and parallel robots have dominated until today, is witnessing a new type of robot named hybrid robots. But it must be admitted that hybrid robots are still in their infancy. More new, hybrid robot designs have begun to attract the attention of the robot community. To contribute to this emerging interest in hybrid robots, a new planar hybrid robot is proposed in this book chapter. First, the general structure and conceptual design of this new planar hybrid robot, including joint types, are explained in detail. Following this, the schematic of the robot is given. Second, the kinematic equations of this robot are obtained using the DH method. Afterward, the inverse kinematics of the proposed hybrid robot is obtained. Finally, a numerical example is given to better understand the proposed new planar parallel robot.

Keywords: inverse kinematics, hybrid robots, serial robots, parallel robots, grippers, DH parameters

1. Introduction

In general, robot manipulators can be classified as serial robots [1, 2], parallel robots [3, 4], and hybrid robots [5, 6]. Serial robot manipulators were the first type of robot to be studied, and these studies began in the early 1950s. Unimate, patented by American inventor George Devol in 1954, was the first industrial robot designed. In addition, the Unimate 1900 series was the first mass-production robotic arm devoted to factory automation. Later, highly advanced industrial robots were produced in developed countries of the world. For example, ABB IRB 1660ID is a high-performance serial robot devoted to arc welding and machine tending. KUKA KR AGILUS' small robot family is extremely fast, especially when picking and placing tasks. They have high repeatability and precision. They have large accessible areas such as access near the robot base and overhead areas. When parallel robots are considered, General Stewart platform manipulator [7] and delta parallel robot [8] are the most prominent parallel robots. For example, Kawasaki YF003N delta parallel is good at high-speed pick and place applications such as pick and place foods and chopsticks on a Conveyor. H-850 6-axis hexapod (Stewart platform) produced by PI company is good at high load positioning. It has high precision performance with large rotation and translation capabilities. It is possible to multiply these examples for serial and parallel robots. However, in terms of hybrid robots, the situation is not the same as in serial and parallel robots. There are currently very few hybrid robots available. Almost all of them are at the design stage [5, 6]. One of the rare robots that has reached the process of being used in industry is the 5-DOF hybrid robot [9]. The 3-DOF is used for the general positioning of the camera, and the remaining 2-DOF is used for orientation. The most important benefit of the 5-DOF hybrid robot is the accessibility performance. In other words, reaching areas that cannot be reached with classical instruments is the most important benefit of this robot [9]. Therefore, the 5-DOF parallel hybrid robot is used for minimally invasive surgery.

Serial and parallel robots have some advantages and disadvantages compared to each other such as high precision, accessibility, rigidity, and stiffness [4, 10]. Hybrid manipulators are potential candidates for problems that serial and parallel robots are not able to overcome. A hybrid robot manipulator can be described as a combination of serial and parallel robot manipulators or a combination of two parallel manipulators [5]. As mentioned before, studies on hybrid parallel robots are still in their infancy. These studies generally focus on spatial hybrid robots [11–14]. A new planar hybrid robot manipulator is proposed in this book chapter. Firstly, the general structure of this planar hybrid robot is explained in detail. Secondly, the inverse kinematics of this robot is performed by using DH method [15], which is a well-known method in robotic research. Finally, a numerical example is provided to analyze the proposed new planar parallel robot in a more comprehensive approach.

2. The structure of the new planar hybrid robot manipulators

The new hybrid robot structure is composed of one 2-DOF planar parallel robot manipulator and two planar serial robot manipulators, as shown in **Figure 1**. Two 2-DOF planar serial manipulators are mounted on 2-RPR planar parallel manipulators, as illustrated in **Figure 1**, where the new structure constitutes a hybrid robot manipulator [5]. Two end-effectors of the 1-DOF planar serial manipulators and 2-RPR planar parallel manipulators work altogether as cooperating manipulators [5]. The 1-DOF planar serial manipulators also have the layout angles ϑ_i that design the direction of the end-effector. The points P_{ef} and P_{mp} define the end-effector positions of hybrid manipulator and intermediate moving platform, respectively.

The points E_1 and E_2 illustrates the end points of each serial planar manipulators. Each of the 1-DOF planar serial manipulators has Revolute (R) joint structure where the joints are denoted as ϕ_1 and ϕ_2, and the layout angles threated as second revolute joints are represented by the symbols ϑ_1 and ϑ_2. As it is well-known that all of the revolute joints (ϕ_1, ϕ_2) of two1-DOF serial planar parallel robot manipulators are active joints. However, not all of the joints are active for parallel robot manipulators. Each of the 2-DOF planar parallel robot manipulators has Revolute-Prismatic (RP) joint structures where the first passive revolute joints are denoted as θ_1 and θ_2, and the second active prismatic joints are represented by the symbols d_1 and d_2. The link lengths are illustrated by the symbols l_i, h_i and f_i where $i = 1, 2$. The symbol γ illustrates the orientation of the intermediate moving platform attached between the points M_1 and M_2. The distances on **Figure 1** are described as $P_{mp}M_1 = r_{m_1}$, $P_{mp}M_2 = r_{m_2}$, $P_{mp}C_1 = r_{s_1}$, $P_{mp}C_2 = r_{s_2}$. $E_1P_{ef} = r_1$ and $E_2P_{ef} = r_2$. The base and moving frame coordinates of the habit robot are described as $B_1 = (b_{x_1}, b_{y_1})$, $B_2 = (b_{x_2}, b_{y_2})$ and $M_1 = (m_{x_1}, m_{y_1})$, $M_2 = \left(m_{x_3}, m_{y_3}\right)$, respectively. The lines z_1 and z_2 are the zero

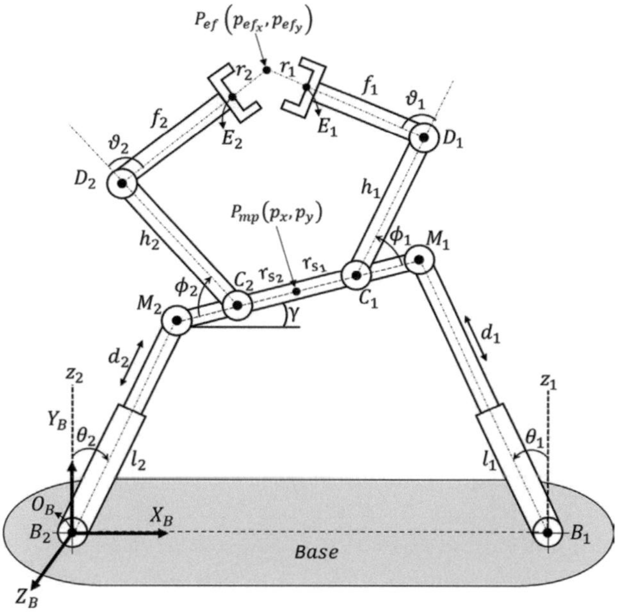

Figure 1.
A new planar hybrid robot manipulator.

position lines of the parallel robot. The passive joint angles rotates with respect to this lines. The revolute joints rotates with respect to the moving platform line M_1M_2.

3. The kinematic equations of the planar hybrid robot manipulator

A few different loop closure equations can be written to describe the kinematics of the manipulator from the base frame to the end-effector. One of the loop closure equations of the manipulator, considering **Figure 1**, can be written as

$$O_BP_{ef} = O_BP_{mp} + P_{mp}C_i + C_iD_i + D_iP_{ef} \quad i = 1, 2. \tag{1}$$

The right-hand side of the Eq. (1) can be written as

$$
\begin{aligned}
{}^{O_B}_{P_{ef}}T^i_R &= {}^{O_B}_{P_{mp}}T\,{}^{P_{mp}}_{C_i}T\,{}^{C_i}_{D_i}T\,{}^{D_i}_{P_{ef}}T \quad i = 1, 2. \\
&= \begin{bmatrix} c\rho_i & -s\rho_i & 0 & p_x \\ s\rho_i & c\rho_i & 0 & p_y \\ 0 & 0 & 1 & 0 \\ 0 & 0 & 0 & 1 \end{bmatrix}
\begin{bmatrix} c\phi_i & -s\phi_i & 0 & r_{s_i} \\ s\phi_i & c\phi_i & 0 & 0 \\ 0 & 0 & 1 & 0 \\ 0 & 0 & 0 & 1 \end{bmatrix}
\begin{bmatrix} c\vartheta_i & -s\vartheta_i & 0 & h_i \\ s\vartheta_i & c\vartheta_i & 0 & 0 \\ 0 & 0 & 1 & 0 \\ 0 & 0 & 0 & 1 \end{bmatrix}
\begin{bmatrix} 1 & 0 & 0 & f_{r_i} \\ 0 & 1 & 0 & 0 \\ 0 & 0 & 1 & 0 \\ 0 & 0 & 0 & 1 \end{bmatrix}
\end{aligned} \tag{2}
$$

where $\rho_1 = \gamma, \rho_2 = \pi + \gamma$ and $f_{r_i} = f_i + r_i$ The left-hand side of the Eq. (1) can be written as

$${}^{O_B}_{P_{ef}}T^i_L = \begin{bmatrix} r_{11} & r_{12} & 0 & p_{ef_x} \\ r_{21} & r_{22} & 0 & p_{ef_y} \\ 0 & 0 & 1 & 0 \\ 0 & 0 & 0 & 1 \end{bmatrix} \quad i = 1, 2 \tag{3}$$

The following equation can be written since position vectors of ${}^{O_B}_{P_{ef}}T^i_L$ and ${}^{O_B}_{P_{ef}}T^i_R$ are equal. Note that after performing multiplications, only the position vectors in Eq. (4), Eq. (5), and Eqs. (6)–(8) are equal. Rotation matrices are not equal and not used for finding the invers kinematics of a new planar hybrid robot manipulator.

$$\begin{bmatrix} c\rho_i & -s\rho_i & 0 & p_x \\ s\rho_i & c\rho_i & 0 & p_y \\ 0 & 0 & 1 & 0 \\ 0 & 0 & 0 & 1 \end{bmatrix} \begin{bmatrix} c\phi_i & -s\phi_i & 0 & r_{s_i} \\ s\phi_i & c\phi_i & 0 & 0 \\ 0 & 0 & 1 & 0 \\ 0 & 0 & 0 & 1 \end{bmatrix} \begin{bmatrix} c\vartheta_i & -s\vartheta_i & 0 & h_i \\ s\vartheta_i & c\vartheta_i & 0 & 0 \\ 0 & 0 & 1 & 0 \\ 0 & 0 & 0 & 1 \end{bmatrix} \begin{bmatrix} 1 & 0 & 0 & f_{r_i} \\ 0 & 1 & 0 & 0 \\ 0 & 0 & 1 & 0 \\ 0 & 0 & 0 & 1 \end{bmatrix} = \begin{bmatrix} 1 & 0 & 0 & p_{ef_x} \\ 0 & 1 & 0 & p_{ef_y} \\ 0 & 0 & 1 & 0 \\ 0 & 0 & 0 & 1 \end{bmatrix} \quad i = 1,2 \qquad (4)$$

Multiplying both sides of Eq. (4) by inverse of ${}^{O_B}_{P_{mp}}T\,{}^{P_{mp}}_{C_i}T$ yields

$$\begin{bmatrix} c\vartheta_i & -s\vartheta_i & 0 & h_i \\ s\vartheta_i & c\vartheta_i & 0 & 0 \\ 0 & 0 & 1 & 0 \\ 0 & 0 & 0 & 1 \end{bmatrix} \begin{bmatrix} 1 & 0 & 0 & f_{r_i} \\ 0 & 1 & 0 & 0 \\ 0 & 0 & 1 & 0 \\ 0 & 0 & 0 & 1 \end{bmatrix} \quad i = 1,2$$
$$= \begin{bmatrix} c(\rho_i+\phi_i) & s(\rho_i+\phi_i) & 0 & -p_x c(\rho_i+\phi_i) - p_y s(\rho_i+\phi_i) - r_{s_i} c\phi_i \\ -s(\rho_i+\phi_i) & c(\rho_i+\phi_i) & 0 & p_x s(\rho_i+\phi_i) - p_y c(\rho_i+\phi_i) + r_{s_i} s\phi_i \\ 0 & 0 & 1 & 0 \\ 0 & 0 & 0 & 1 \end{bmatrix} \begin{bmatrix} 1 & 0 & 0 & p_{ef_x} \\ 0 & 1 & 0 & p_{ef_y} \\ 0 & 0 & 1 & 0 \\ 0 & 0 & 0 & 1 \end{bmatrix} \qquad (5)$$

Performing matrix multiplications produces the following equation. Since only position vectors of transformation matrices on Eq. (5) are equal to each other, the elements of rotation matrices of both sides of Eqs. (6)–(8) are represented by dots:

$$\begin{bmatrix} . & . & . & h_i + f_{r_i} c\vartheta_i \\ . & . & . & f_{r_i} s\vartheta_i \\ . & . & . & 0 \\ 0 & 0 & 0 & 1 \end{bmatrix} = \begin{bmatrix} . & . & . & a_x \\ . & . & . & a_y \\ . & . & . & 0 \\ 0 & 0 & 0 & 1 \end{bmatrix} \quad i = 1,2 \qquad (6)$$

where

$$\begin{aligned} a_x = {} & \left[\left(p_{ef_x} - p_x\right) c\rho_i + \left(p_{ef_y} - p_y\right) s\rho_i - r_{s_i}\right] c\phi_i \\ & + \left[\left(p_{ef_y} - p_y\right) c\rho_i - \left(p_{ef_x} - p_x\right) s\rho_i\right] s\phi_i \end{aligned} \quad i = 1,2 \qquad (7)$$

$$\begin{aligned} a_y = &\left[-\left(p_{ef_x} - p_x\right)c\rho_i - \left(p_{ef_y} - p_y\right)s\rho_i + r_{s_i}\right]s\phi_i \\ &+\left[\left(p_{ef_y} - p_y\right)c\rho_i - \left(p_{ef_x} - p_x\right)s\rho_i\right]c\phi_i \end{aligned} \quad i = 1,2 \tag{8}$$

Since only position vectors are equal let us write position vector as

$$\begin{bmatrix} h_i + f_{r_i}c\vartheta_i \\ f_{r_i}s\vartheta_i \\ 0 \end{bmatrix} = \begin{bmatrix} a_x \\ a_y \\ 0 \end{bmatrix} \quad i = 1,2 \tag{9}$$

Squaring both sides of Eq. (9) yields

$$h_i^2 + 2h_if_{r_i}c\vartheta_i + f_{r_i}^2 = a_i \;\; i = 1,2 \tag{10}$$

where

$$\begin{aligned} a_i = &2r_{s_i}\left[\left(p_x - p_{ef_x}\right)c\rho_i + \left(p_y - p_{ef_y}\right)s\rho_i\right] \\ &+r_{s_i}^2 + p_x^2 + p_y^2 + p_{ef_x}^2 + p_{ef_y}^2 - h_i^2 - f_{r_i}^2 - 2\left(p_xp_{ef_x} + p_yp_{ef_y}\right) \end{aligned} \quad i = 1,2 \tag{11}$$

Extracting $c\vartheta_i$ produces

$$c\vartheta_i = \frac{a_i - h_i^2 - f_{r_i}^2}{2h_if_{r_i}} \;\; i = 1,2 \tag{12}$$

The first joint is found as

$$\vartheta_i = Atan2\left(\sqrt{1 - \left(\frac{a_i - h_i^2 - f_{r_i}^2}{2h_if_{r_i}}\right)^2}, \frac{a_i - h_i^2 - f_{r_i}^2}{2h_if_{r_i}}\right) \;\; i = 1,2 \tag{13}$$

Equating (1, 4) matrix element of each side of Eqs. (6)–(8) yields

$$A_is\phi_i + B_ic\phi_i = C_i \;\; i = 1,2 \tag{14}$$

where

$$\begin{aligned} A_i &= \left(p_{ef_y} - p_y\right)c\rho_i - \left(p_{ef_x} - p_x\right)s\rho_i \\ B_i &= \left(p_{ef_x} - p_x\right)c\rho_i + \left(p_{ef_y} - p_y\right)s\rho_i - r_{s_i} \\ C_i &= h_i + f_{r_i}c\vartheta_i \end{aligned} \quad i = 1,2 \tag{15}$$

The first joint is found as

$$\phi_i = Atan2(A_i, B_i) \pm Atan2\left(\sqrt{A_i^2 + B_i^2 - C_i^2}, C_i\right) \;\; i = 1,2 \tag{16}$$

The other loop closure equations of the manipulator considering **Figure 1** can be written as

$$O_BB_i + B_iM_i = O_BP_{mp} + P_{mp}M_i \quad i = 1,2 \tag{17}$$

The left-hand side of the Eq. (17) can be written as

$$
\begin{aligned}
{}^{OB_i}_{B_iM_i}T_L &= {}^{O_B}_{B_i}T\,{}^{B_i}_{M_i}T \\
&= \begin{bmatrix} 1 & 0 & 0 & b_{x_i} \\ 0 & 1 & 0 & b_{y_i} \\ 0 & 0 & 1 & 0 \\ 0 & 0 & 0 & 1 \end{bmatrix}
\begin{bmatrix} c\theta_i & -s\theta_i & 0 & 0 \\ s\theta_i & c\theta_i & 0 & 0 \\ 0 & 0 & 1 & 0 \\ 0 & 0 & 0 & 1 \end{bmatrix}
\begin{bmatrix} 1 & 0 & 0 & 0 \\ 0 & 0 & 1 & l_i + d_i \\ 0 & -1 & 0 & 0 \\ 0 & 0 & 0 & 1 \end{bmatrix} \quad i = 1,2 \\
&= \begin{bmatrix} c\theta_i & 0 & -s\theta_i & b_{x_i} - s\theta_i(l_i + d_i) \\ s\theta_i & 0 & c\theta_i & b_{y_i} + c\theta_i(l_i + d_i) \\ 0 & -1 & 0 & 0 \\ 0 & 0 & 0 & 1 \end{bmatrix}
\end{aligned} \tag{18}
$$

The left-hand side of the Eq. (17) can be written as

$$
\begin{aligned}
{}^{OB_i}_{B_iM_i}T_R &= {}^{O_B}_{P_{mp}}T\,{}^{P_{mp}}_{M_i}T \\
&= \begin{bmatrix} c\rho_i & -s\rho_i & 0 & p_x \\ s\rho_i & c\rho_i & 0 & p_y \\ 0 & 0 & 1 & 0 \\ 0 & 0 & 0 & 1 \end{bmatrix}
\begin{bmatrix} 1 & 0 & 0 & r_{m_i} \\ 0 & 1 & 0 & 0 \\ 0 & 0 & 1 & 0 \\ 0 & 0 & 0 & 1 \end{bmatrix}
= \begin{bmatrix} c\rho_i & -s\rho_i & 0 & m_{x_i} \\ s\rho_i & c\rho_i & 0 & m_{y_i} \\ 0 & 0 & 1 & 0 \\ 0 & 0 & 0 & 1 \end{bmatrix} \quad i = 1,2
\end{aligned} \tag{19}
$$

where

$$m_{x_i} = p_x + r_{m_i}c\rho_i \quad m_{y_i} = p_y + r_{m_i}s\rho_i \quad i = 1,2 \tag{20}$$

Since the position vectors of ${}^{OB_i}_{B_iM_i}T_L$ and ${}^{OB_i}_{B_iM_i}T_R$ are equal, the following relation can easily be written. Note that after performing multiplications, only the position vectors in Eq. (21), Eq. (22), and Eq. (23) are equal. Rotation matrices are not equal and not used for finding the invers kinematics of a new planar hybrid robot manipulator.

$$
\begin{bmatrix} 1 & 0 & 0 & b_{x_i} \\ 0 & 1 & 0 & b_{y_i} \\ 0 & 0 & 1 & 0 \\ 0 & 0 & 0 & 1 \end{bmatrix}
\begin{bmatrix} c\theta_i & -s\theta_i & 0 & 0 \\ s\theta_i & c\theta_i & 0 & 0 \\ 0 & 0 & 1 & 0 \\ 0 & 0 & 0 & 1 \end{bmatrix}
\begin{bmatrix} 1 & 0 & 0 & 0 \\ 0 & 0 & 1 & l_i + d_i \\ 0 & -1 & 0 & 0 \\ 0 & 0 & 0 & 1 \end{bmatrix}
= \begin{bmatrix} c\rho_i & -s\rho_i & 0 & m_{x_i} \\ s\rho_i & c\rho_i & 0 & m_{y_i} \\ 0 & 0 & 1 & 0 \\ 0 & 0 & 0 & 1 \end{bmatrix} \quad i = 1,2 \tag{21}
$$

Multiplying both sides of Eq. (21) by inverse of ${}^0_1T{}^1_2T$ yields

$$\begin{bmatrix} 1 & 0 & 0 & 0 \\ 0 & 0 & 1 & l_i + d_i \\ 0 & -1 & 0 & 0 \\ 0 & 0 & 0 & 1 \end{bmatrix} = \begin{bmatrix} c\theta_i & s\theta_i & 0 & -b_{x_i}c\theta_i - b_{y_i}s\theta_i \\ -s\theta_i & c\theta_i & 0 & b_{x_i}s\theta_i - b_{y_i}c\theta_i \\ 0 & 0 & 1 & 0 \\ 0 & 0 & 0 & 1 \end{bmatrix} \begin{bmatrix} c\rho_i & -s\rho_i & 0 & m_{x_i} \\ s\rho_i & c\rho_i & 0 & m_{y_i} \\ 0 & 0 & 1 & 0 \\ 0 & 0 & 0 & 1 \end{bmatrix} \quad i = 1,2 \tag{22}$$

Multiplying transformation matrices on the right-hand side of Eq. (22) produces Eq. (23). Since only position vectors of transformation matrices on Eq. (22) are equal to each other, the elements of rotation matrices of both sides of Eq. (23) are represented by dots.

$$\begin{bmatrix} \cdot & \cdot & \cdot & 0 \\ \cdot & \cdot & \cdot & l_i + d_i \\ \cdot & \cdot & \cdot & 0 \\ 0 & 0 & 0 & 1 \end{bmatrix} = \begin{bmatrix} \cdot & \cdot & \cdot & (m_{x_i} - b_{x_i})c\theta_i + (m_{y_i} - b_{y_i})s\theta_i \\ \cdot & \cdot & \cdot & -(m_{x_i} - b_{x_i})s\theta_i + (m_{y_i} - b_{y_i})c\theta_i \\ \cdot & \cdot & \cdot & 0 \\ 0 & 0 & 0 & 1 \end{bmatrix} \quad i = 1,2 \tag{23}$$

After simplification, the active joint variable d_i can be found as

$$d_i = \sqrt{(m_{x_i} - b_{x_i})^2 + (m_{y_i} - b_{y_i})^2} - l_i \qquad i = 1,2 \tag{24}$$

Once the active joint variable d_i is found, the passive joint variable θ_i can be found using Eq. (23). The following equation can be written from Eq. (23).

$$(m_{x_i} - b_{x_i})c\theta_i + (m_{y_i} - b_{y_i})s\theta_i = 0 \qquad i = 1,2 \tag{25}$$

Finally, the passive joint variable θ_i can be found as

$$\theta_i = Atan2(-(m_{x_i} - b_{x_i}), (m_{y_i} - b_{y_i})) \qquad i = 1,2 \tag{26}$$

4. Numerical example

Consider the planar hybrid robot manipulator shown in **Figure 1**. A cylindrical load is attached by the end-effectors of the two serial planar robot manipulators, as shown in **Figure 2**. The cooperative end-effectors of the serial robot manipulators hold the cylindrical load tightly. The symbols r_i denotes the distance between P_{ef} and E_i where $i = 1,2$. The lines $P_{ef}E_i$ and E_iD_i lie on the same line as in **Figure 1**. According to the kinematic parameters given in **Table 1**, let us find active and passive joint variables for three different end-effector poses and draw the manipulator configurations for the three poses. Note that although the layout angles ϑ_1 and ϑ_2 can be designed in advance, they are treated as unknown second revolute joints in this example.

Let us first find the active and passive joint variables for the first end-effector poses as an example. The layout angles ϑ_1 and ϑ_2 can be found by using Eq. (13) as

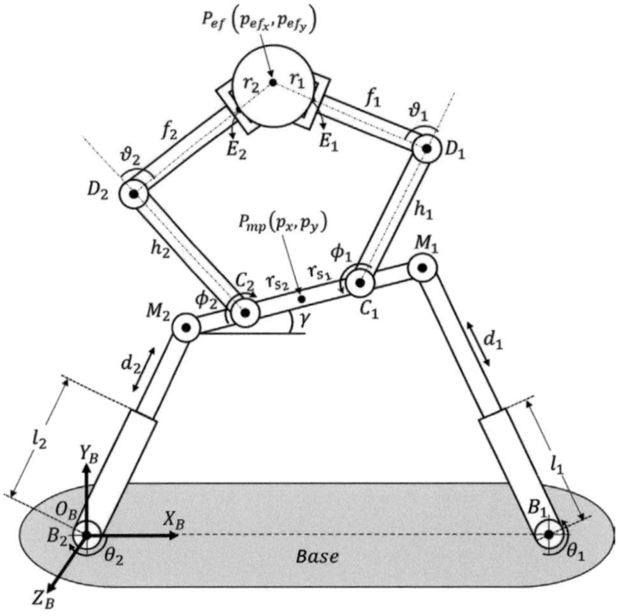

Figure 2.
The schematic of the new hybrid robot manipulator for the numerical example.

Pose	Link lengths	Base coordinates				Lengths in moving			Poses				
									Intermediate pose (P_{mp})			End-effector position (p_{ef})	
	l_i, h_i, f_i	b_{x_1}	b_{y_1}	b_{x_2}	b_{y_2}	r_{m_i}	r_{s_i}	r_i	p_x	p_y	γ	p_{ef_x}	p_{ef_y}
1	$0.6\,m$	$1.4\,m$	0	0	0	$0.3\,m$	$0.2\,m$	$0.1\,m$	$0.75\,m$	$0.8\,m$	30°	$0.5\,m$	$1.91\,m$
2									$0.7\,m$	$1\,m$	0°	$0.7\,m$	$1.90\,m$
3									$0.75\,m$	$0.8\,m$	-30°	$1.05\,m$	$1.90\,m$

Table 1.
Kinematic parameters of planar hybrid parallel manipulator where i = 1, 2.

$$\begin{aligned}
\vartheta_1 &= atan2\left(\sqrt{1-\left(\frac{a_1-h_1^2-f_{r_1}^2}{2h_1f_{r_1}}\right)^2}, \frac{a_1-h_1^2-f_{r_1}^2}{2h_1f_{r_1}}\right) \\
&= atan2\left(\sqrt{1-\left(\frac{1.1992-0.6^2-0.7^2}{2\cdot 0.6\cdot 0.7}\right)^2}, \left(\frac{1.1992-0.6^2-0.7^2}{2\cdot 0.6\cdot 0.7}\right)\right) \\
&= 65.4355^o
\end{aligned} \tag{27}$$

$$\vartheta_2 = -42.4308^o$$

where

$$\begin{aligned} a_1 &= 2r_{s_1}\left[\left(p_x - p_{ef_x}\right)c\rho_1 + \left(p_y - p_{ef_y}\right)s\rho_1\right] + r_{s_1}^2 + p_x^2 + p_y^2 + p_{ef_x}^2 + p_{ef_y}^2 - 2\left(p_x p_{ef_x} + p_y p_{ef_y}\right) \\ &= 2 \cdot 0.2 \cdot [(0.75 - 0.5)\cos(30^o) + (0.8 - 1.91)\sin(30^o)] + 0.2^2 + 0.75^2 + 0.8^2 \\ &+ 0.5^2 + 1.91^2 - 2 \cdot (0.75 \cdot 0.5 + 0.8 \cdot 1.91) \\ &= 1.1992 \end{aligned} \tag{28}$$

The layout angles ϕ_1 and ϕ_2 can be found by using Eq. (16) as

$$\begin{aligned} \phi_1 &= atan2(A_1, B_1) + atan2\left(\sqrt{A_1^2 + B_1^2 - C_i^2}, C_1\right) \\ &= atan2(1.0863, 0.1385) - atan2\left(\sqrt{1.0863^2 + 0.1385^2 - 0.891^2}, 0.891\right) \\ &= 47.1876^2 \end{aligned} \tag{29}$$

$$\phi_2 = -93.4427^o$$

where

$$\begin{aligned} A_1 &= \left(p_{ef_y} - p_y\right)c\rho_i - \left(p_{ef_x} - p_x\right)s\rho_i \\ &= (1.91 - 0.8)\cos(30^o) - (0.5 - 0.75)\sin(30^o) = 1.0863 \\ B_1 &= \left(p_{ef_x} - p_x\right)c\rho_1 + \left(p_{ef_y} - p_y\right)s\rho_1 - r_{s_1} \\ &= (0.5 - 0.75)\cos(30^o) + (1.91 - 0.8)\sin(30^o) - 0.2 = 0.1385 \\ C_1 &= h_1 + f_{r_1}c\vartheta_1 \\ &= 0.6 + 0.7 \cdot \cos(65.4355^o) = 0.891 \end{aligned} \tag{30}$$

The active joint variables d_1 and d_2 can be found by using Eq. (24) as

$$\begin{aligned} d_1 &= \sqrt{\left(m_{x_1} - b_{x_1}\right)^2 + \left(m_{y_1} - b_{y_1}\right)^2} - l_1 \\ &= \sqrt{(1.0098 - 1.4)^2 + (0.95 - 0)^2} - 0.6 = 0.4270m \end{aligned} \tag{31}$$

$$d_2 = 0.2141$$

The passive joint variables θ_1 and θ_2 can be found by using Eq. (26) as

$$\begin{aligned} \theta_1 &= atan2\left(-\left(m_{x_1} - b_{x_1}\right), \left(m_{y_1} - b_{y_1}\right)\right) \\ &= atan2(-(1.0098 - 1.4), (0.95 - 0)) = 22.3294^o \end{aligned} \quad i = 1, 2 \tag{32}$$

$$\theta_1 = -37.0215^o$$

where

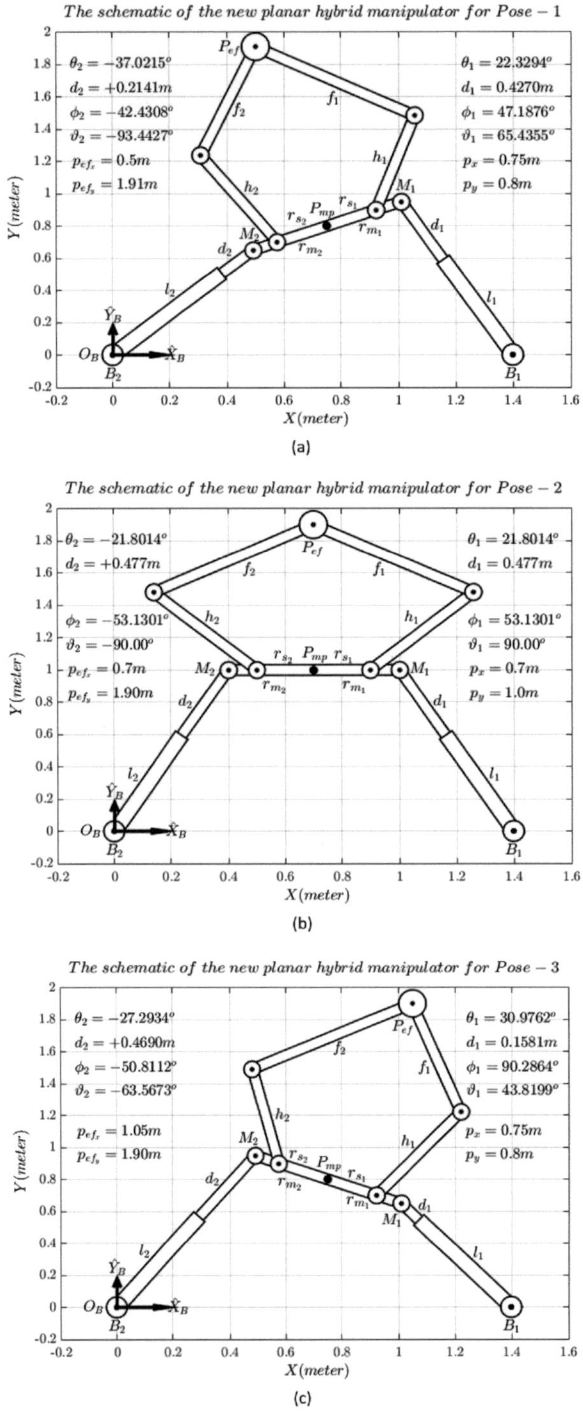

Figure 3.
The three manipulator configurations for pose-1, pose-2 and pose-3.

$$\begin{aligned} m_{x_1} &= p_x + r_{m_1} c\rho_1 & m_{y_1} &= p_y + r_{m_1} s\rho_1 \\ &= 0.75 + 0.3 \cdot \cos(30^o) & &= 0.8 + 0.3 \cdot \sin(30^o) \quad i = 1, 2 \\ &= 1.0098 & &= 0.95 \end{aligned} \tag{33}$$

Figure 3 demonstrates the three manipulator configurations for pose-1, pose-2 and pose-3. The pose of the intermediate moving platform and the end effector, as well as the active and passive joint variables, are shown in the figures to ease the inverse kinematics analyses.

5. Conclusions

A new planar hybrid robot is proposed in this book chapter. The general structure and conceptual design of this new planar hybrid robot are described in detail. Afterward, the inverse kinematics of the proposed hybrid robot is obtained by using the transformation matrices obtained from DH method. Finally, a numerical example is given for the three manipulator configurations.

The planar hybrid robot manipulator proposed here has a design open to improvement. The two serial robot manipulators that hold the objects work like a gripper. Materials can be held more easily by changing the lengths and shapes of the f_1 and f_2 links of these serial robots. Additionally, in this example, considering that layout angles ϑ_1 and ϑ_2 are unknown, the inverse kinematics is applied to find these angles. Some important issues can be addressed by giving different values to the layout angles in advance. For example, the grasping force of objects can be reduced by optimizing the layout angles ϑ_1 and ϑ_2. Thus, the energy consumed by the motors can be reduced during the gripping process. Another issue is that serial robots can be used for medical purposes. Each serial robot can undertake different tasks. While one of the serial robots holds the sewing thread, the other can cut this thread. As a result, the proposed planar hybrid robot has a design that is open to development and a structure that is open to versatile use.

Author details

Serdar Küçük
Technology Faculty, Biomedical Engineering, Kocaeli University, Umuttepe Kampus, Kocaeli, Türkiye

*Address all correspondence to: skucuk@kocaeli.edu.tr

References

[1] Ayiz C, Kucuk S. The kinematics of industrial robot manipulators based on the exponential rotational matrices. In: IEEE International Symposium on Industrial Electronics (ISIE 2009), Seoul Olympic Parktel, Seoul, Korea, July 5-8, 2009. IEEE; 2009. pp. 977-982

[2] Toz M, Kucuk S. A comparative study for computational cost of fundamental robot manipulators. In: IEEE International Conference on Industrial Technology. IEEE; 2011. pp. 289-293

[3] Toz M, Kucuk S. Development of derivation of inverse Jacobian matrices for 195 6-DOF GSP mechanisms. Turkish Journal of Electrical Engineering & Computer Sciences. 2016;**24**:4142-4153

[4] Sun T, Lian B. Stiffness and mass optimization of parallel kinematic machine. Mechanism and Machine Theory. 2018;**120**:73-88

[5] Tanev TK. Kinematics of a hybrid (parallel-serial) robot manipulator. Mechanism and Machine Theory. 2000; **35**(9):1183-1196

[6] Kucuk S. Dexterous workspace optimization for a new hybrid parallel robot manipulator. Journal of Mechanisms and Robotics. 2018;**10**(6): 064503

[7] Stewart D. A platform with six degrees of freedom. Proceedings of the Institution of Mechanical Engineers, Part C. 1965;**180**(1):371-386

[8] Clavel R. Device for the Movement and Positioning of an Element in Space. Patent, US 4976582; 1990

[9] Pisla D, Gherman B, Plitea N, Gyurka B, Calin V, Vlad L, et al. PARASURG hybrid parallel robot for minimally invasive surgery. Chirurgia. 2011;**5**(4):629-625

[10] Giberti H, Ferrari D. A novel hardware-in-the-loop device for floating offshore wind turbines and sailing boats. Mechanism and Machine Theory. 2015; **85**:82-105

[11] Antonov AV, Fomin AS. Inverse kinematics of a 5-DOF hybrid manipulator. Automation and Remote Control. 2023;**84**(3):281-293

[12] Antonov A, Fomin A, Glazunov V, Kiselev S, Carbone G. Inverse and forward kinematics and workspace analysis of a novel 5-DOF (3T2R) parallel-serial (hybrid) manipulator. International Journal of Advanced Robotic Systems. 2021;**18**(2):2963. DOI: 10.1177/1729881421992963

[13] Ye H, Wang D, Wu J, Yue Y, Zhou Y. Forward and inverse kinematics of a 5-DOF hybrid robot for composite material machining. Robotics and Computer-Integrated Manufacturing. 2020;**65**:101961. DOI: 10.1016/j.rcim.2020.101961

[14] Hu B, Shi Y, Xu L, Bai P. Reconsideration of terminal constraint/mobility and kinematics of 5-DOF hybrid manipulators formed by one 2R1T PM and one RR SM. Mechanism and Machine Theory. 2020;**149**:103837. DOI: 10.1016/j.mechmachtheory.2020.103837

[15] Denavit J, Hartenberg RS. A kinematic notation for lower-pair mechanisms based on matrices. Journal of Applied Mechanics. 1955;**1**:215-221

Chapter 5

Kinematic Modeling of Continuum Robots Using Artificial Intelligence Tools

Abdelhamid Ghoul, Selman Djeffal and Kamel Kara

Abstract

Continuum robots are progressively dwarfing traditional rigid robots in many areas, due to their outstanding flexibility and easy adaptation to almost any kind of trajectory, yet their kinematic modeling is presented as a drawback, in particular with multi-sections and variable curvature, which curbs the full benefit of these robots. In this context, this chapter presents the description of continuum robots and their kinematic modeling in detail. Firstly, the description of the behavior and forward kinematic models are presented for continuum robots with constant and variable curvatures. Then, to calculate the inverse kinematic model of continuum robots, artificial neural networks, and metaheuristic optimization techniques have been used to obtain the inverse kinematic model of single and multi-section continuum robots, respectively. To validate the utility of the proposed techniques, different trajectories are considered to be followed by considered robots. It is found that the developed techniques are powerful tools to deal with the high complexity of nonlinear equations, in particular when it comes to solving the inverse kinematic model of multi-section continuum robots with variable curvature. To have a closer look at the efficiency of the developed techniques during the follow-up of the proposed trajectories, 3D simulations have been carried out with different configurations.

Keywords: continuum robots, kinematic modeling, artificial neural networks, metaheuristic optimization, 3D animations

1. Introduction

Robotics increasingly attract researchers' interest over the past few decades, as it has become the solution to many problems in real world applications. Especially, when it comes to difficult and complex tasks that humans cannot handle or in industrial environments that may be hazardous to humans (for example near high temperatures and nuclear reactors or when dealing with chemical waste and sharp objects). As a consequence, the field of robotics has attracted more researchers to design more

IntechOpen

sophisticated robots to handle the required tasks. Initially, rigid robots were developed to perform specific tasks, for example in some military tasks, aerospace exploration and industrial operations because of their dexterity especially in repetitive tasks that require precision and rigidity.

Outside of industry, rigid robots have shown some drawbacks, in particular when it comes to surgery, rescue under buildings, inspection in hazardous environments and nuclear reactors which have labyrinth-like paths and small clustered environments. The deficiency of rigid robots is labeled to their rigid links and their inability to adapt with these kinds of paths. To this end, it would be tempting to find an alternative to these robots. Researchers have developed the so-called continuum bionic robots which can easily be adapted to any kind of paths thanks to their flexibility and high performance. Continuum robots are considered by a large number of researchers as an exciting and new technology because they have notable advantages such as flexibility, lightweight, inherent safety and so forth [1–12]. Continuum robots can also avoid obstacles, handle adjustable stiffness on irregular objects, and even adapt their body shape while maneuvering in an unknown and non-uniform environment. However, they still present an impediment to researchers, because of their complexity when it comes to modeling and control. Despite the fact that many researchers have come up with new methodologies to model them, yet they are not sufficiently reliable and further accuracy is required. Therefore, in this context, this chapter presents a contribution to the kinematics of continuum robots using artificial intelligence tools.

2. Forward kinematic modeling

The Forward Kinematic Model FKM is the set of relationships that allow finding the position of the robot end-effector according to the operating coordinates; unlike rigid robots, the variables that express the configuration of a flexible robot change with respect to the robot's morphology and its type of actuation. For this reason, research on the modeling of flexible continuum robots investigates other means to develop exhaustive kinematic models taking into account maximum of specificities and mechanical properties of the studied robot. To this end, the main objective of this section is to establish the forward kinematic model of a flexible continuum robot with Variable Curvature (VC) and Constant Curvature (CC). First, a general description of the continuum robots is presented by defining the parameters and variables. After that, the forward kinematic modeling problem is formulated and explained. Finally, the workspace is graphically presented based on the established forward kinematic model of continuum robots with both (VC) and (CC).

2.1 Geometric description of cable-driven continuum robot

A cable-driven continuum robot is one of the most commonly used types. Cable-driven continuum robots consist of a flexible backbone with disks that are separated from each other by the same distance, and attached to the backbone.

In most of these robots, the cable is considered as a straight or smooth helical curve. The cables pass through holes that are located on the disks to guide the robot and make it bend to the desired position.

2.1.1 Geometry of the whole profile of the considered robot

Developing a thorough description of the behavior and the morphology of continuum robots are the first major step that is needed to derive precise kinematic models, which themselves help to simulate the movements and the behavior of these robots. Therefore, several models have been proposed by researchers for the geometric description of continuum robots assuming constant curvature [13–20] and variable curvature [21–24].

The kinematics description of the constant and variable curvature continuum robot depicted in **Figure 1** are briefly presented, whose profile is similar to a backbone curve. The complex structure in each continuum robot's has two sections. Each section k is mounted at the base of its first unit, while the remaining units are sequentially stacked on top. Each section has two degrees of freedom, a bending angle $\left(\theta_{j,k}\right)$ and an orientation angle $\left(\varphi_{j,k}\right)$ actuated by three independent cables.

To have a simplified description of the continuum robot under consideration, the profile of the whole section is assimilated into a curve formed by a serial concatenation of circular arcs, (**Figure 2**) which represents the robot central axis [22, 23, 25].

A detailed illustration of a cylindrical and a conical shape units showing arc parameters and geometric properties are discussed in the following.

2.1.2 The geometry of a unit

Grasping the geometry of a single unit paves the way to understanding the remaining units that constitute the backbone of the robot. Each unit consists of two platforms (**Figure 3**), which can be virtual as in the case of CBHA robot or real as in our case. In the general case, the base of the unit is fixed and the upper part is mobile. They are connected by three segments.

For constant curvature continuum robot, each unit of the robot has a lower and an upper disk which have the same diameters (cylindrical shaped unit) and connected through the points $A_{i,j,k}$ and $\widehat{B}_{i,j,k}$. For variable curvature continuum robot, each unit has two disks with different diameters (conical shaped unit) and connected through $B_{i,j,k}$ and $A_{i,j,k}$ (i = 1, 2, 3). Interestingly, the differentiation of the disks' diameters

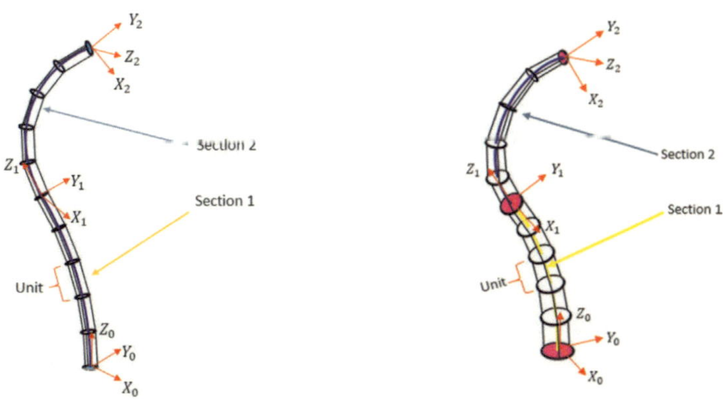

Figure 1.
(left) two-section constant curvature continuum robot with its coordinate frames for each section k*; (right) two-section variable curvature continuum robot with its coordinate frames for each section* k.

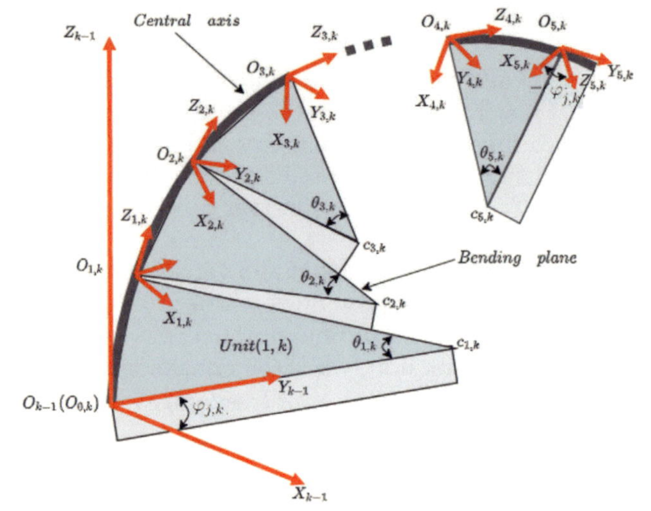

Figure 2.
Frames placements on the central axis for each unit.

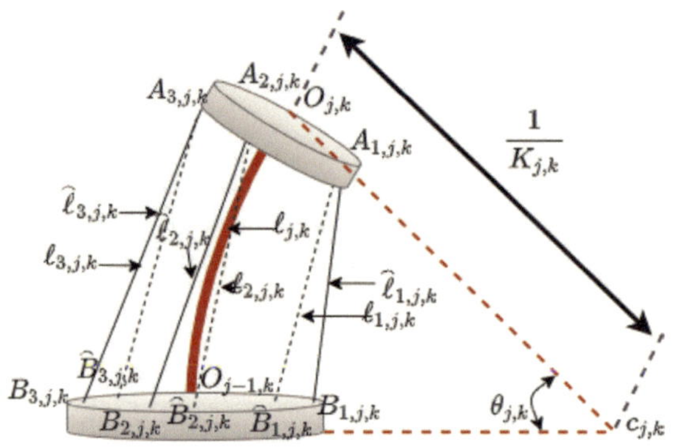

Figure 3.
Description of a conical and cylindrical unit and its cables.

paves the way to establishing an equation which relates the units with each other as it is explained in detail in [22].

The length of cable that connects the disks of a cylindrical unit is denoted by $\ell_{i,j,k}$ and the length of cable that connects the disks of a conical unit is denoted by $\hat{\ell}_{i,j,k}$. Each cylindrical unit is parameterized by its orientation angle $\varphi_{j,k}$, its bending angle $\theta_{j,k}$, and its curvature $\kappa_{j,k}$.

2.1.3 The attached frames on the robot's backbone

In this section, we define the different frames that must be attached to the general geometry of a flexible robot. First, the center of the upper platform of each section k is referenced by a frame named R_k. Then, a frame R_O is fixed at the base of the first section as shown in **Figure 4**.

In order to calculate the FKM of each unit constituting the robot, an intermediate reference frame $R_{j,k}$ is assigned to the center of the platform of each unit (j, k).

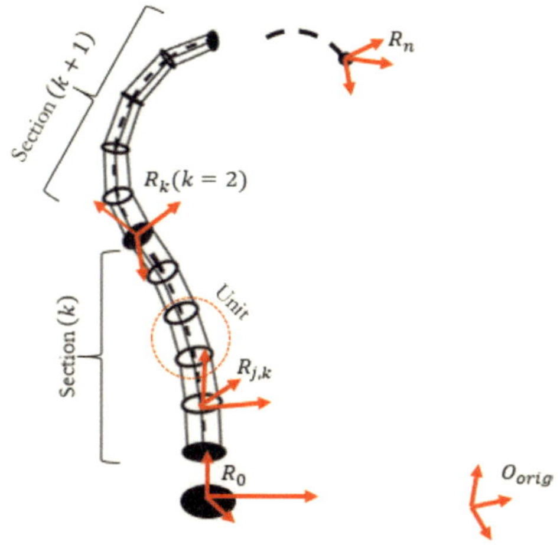

Figure 4.
Placement of frames on the structure of a flexible manipulator.

The description of a flexible section from its base to its end is as follows: the indices of the units j increase from 1 to m, and for each intermediate platform (j, k), the distance between the central axis and the anchor points is noted $r_{j,k}$ (**Figure 3**). The unit (j, k), is modeled as an inextensible arc of a circle. This unit has one end fixed at the origin of the reference frame $O_{j-1,k}$, the other end $O_{j,k}$ is located at a point in the reachable workspace as shown in **Figure 3**.

2.2 Modeling assumptions

Before starting the modeling of the considered robot, it is necessary to point out the different used assumptions, on which the mathematical development throughout this work is based. The previously developed work to model the central axis of flexible continuum robots is based on the constant curvature [26], which considers the deformation of the robot sections as a circular arc. The researchers purposefully used it because it facilitates the calculation of kinematic models. However, the constant curvature does not show a better imitation of the central axis of the continuum robot. For this reason, recent research has opened new doors to modeling the central axis of the continuum robot, which describe in detail the deformation of each unit of the robot as an arc of a circle, typically called variable curvature [21].

The assumptions considered for modeling constant and variable curvature along this chapter can be summarized as follows [22]:

- The flexible continuum robot is described as an open kinematic chain of n sections.
- Each section is a set of equidistant units.
- Each unit is modeled as an inextensible circular arc having its individual parameters.
- The cable lengths are homogeneously fragmented along the robot.

- Robot deformations at sections and units are done without torsion (neglected torsion).

In the last section, a particular case is taken into account where the robot's backbone can be extensible (**Figure 5**).

2.3 Forward kinematic model

The forward kinematic model consists of calculating the position and the orientation of the flexible continuum robot's end-effector as a function of the cable lengths. The FKM for the considered robot in this work can be derived in three steps:

- The FKM of a single unit.
- The FKM of a single section.
- The FKM of a multi-section flexible continuum robot.

2.3.1 FKM of a single unit

As shown in **Figure 3**, an elaborate description of the conical and cylindrical shape of the flexible continuum robot is described.

Overall, the FKM of a unit consists of representing the operational coordinates $X_{j,k}$ as a function of the length of the cables $\widehat{Q}_{j,k}$ through the arc parameters $K_{j,k}$, (See **Figure 5**), and which is obtained by two transformations, the first is the specific transformation and the second is the independent one. Both transformations will be discussed in detail in the rest of this chapter. Assuming that each unit deforms in the form of an arc. The geometric parameters of each unit can be illustrated in a three-

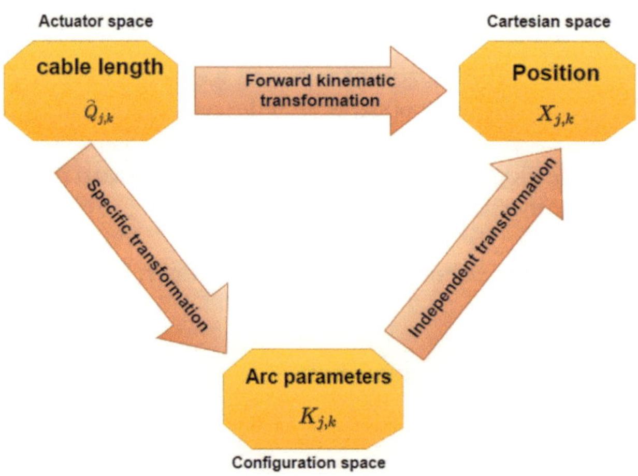

Figure 5.
Global view of the modeling.

dimensional frame as shown in **Figure 6**. The homogeneous transformation can be written by the following equation [18]:

$$T_{j,k}^{j-1,k} = \begin{bmatrix} R_{j,k}^{j-1,k} & O_{j,k}^{j-1,k} \\ 0_{1\times 3} & 1 \end{bmatrix} \quad (1)$$

with $R_{j,k}^{j-1,k}$ and $O_{j,k}^{j-1,k}$ are respectively the (3×3) matrix and the (3×1) vector defining the orientation and the position of the reference frame origin $R_{j,k}$ in the reference frame $R_{j-1,k}$.

Based on the previously defined assumptions, the orientation matrix $R_{j,k}^{j-1,k}$ is given by the following equation (**Figure 6**) [19]:

$$\begin{aligned} R_{j,k}^{j-1,k} &= rot\left(Z_{j-1,k}, \varphi_{j,k}\right) rot\left(Y_{j-1,k}, \theta_{j,k}\right) rot\left(Z_{j-1,k}, -\varphi_{j,k}\right) \\ &= \begin{bmatrix} c^2\varphi_{j,k}c\theta_{j,k} + s^2\varphi_{j,k} & c\varphi_{j,k}c\theta_{j,k}s\varphi_{j,k} - c\varphi_{j,k}s\varphi_{j,k} & c\varphi_k s\theta_{j,k} \\ c\varphi_{j,k}c\theta_{j,k}s\varphi_{j,k} - c\varphi_{j,k}s\varphi_{j,k} & s^2\varphi_{j,k}c\theta_{j,k} + c^2\varphi_{j,k} & s\varphi_{j,k}s\theta_{j,k} \\ -c\varphi_{j,k}s\theta_{j,k} & -s\varphi_{j,k}s\theta_{j,k} & c\theta_{j,k} \end{bmatrix} \end{aligned} \quad (2)$$

with: $c. = \cos(.)$ and $s. = \sin(.)$.
The position vector is defined as follows:

$$O_{j,k}^{j-1,k} = \begin{cases} \dfrac{l_{j,k}}{\theta_{j,k}}\left(1 - \cos\left(\theta_{j,k}\right)\right)\cos\left(\varphi_{j,k}\right) \\ \dfrac{l_{j,k}}{\theta_{j,k}}\left(1 - \cos\left(\theta_{j,k}\right)\right)\sin\left(\varphi_{j,k}\right) \\ \dfrac{l_{j,k}}{\theta_{j,k}}\sin\left(\varphi_{j,k}\right) \end{cases} \quad (3)$$

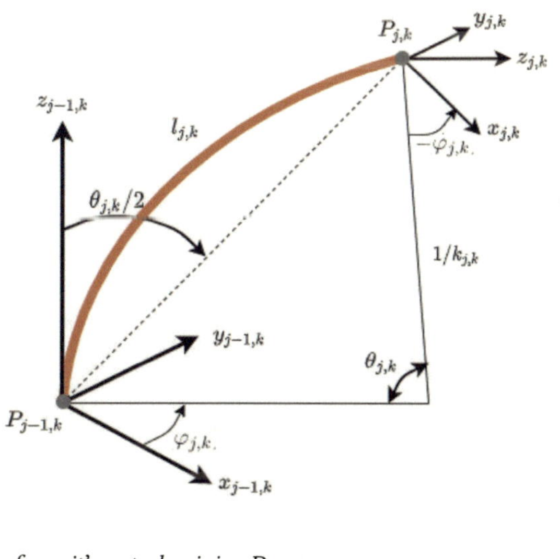

Figure 6.
The geometric parameters of a unit's central axis in 3D space.

2.3.2 Specific transformation

In order to show the length of the cables in the orientation matrix, a specific transformation is established. It expresses the length of the cables $\ell_{i,j,k}$, of each unit (j, k) as a function of the arc parameters which are the orientation angle $\varphi_{j,k}$, the bending angle $\theta_{j,k}$ $[rad]$ and its curvature $\kappa_{j,k}$ $[1/mm]$. First, the orientation angle $\varphi_{j,k}$ must be calculated.

For the variable curvature continuum robot the diameter of each disk is calculated by a general equation that allows determining each of them as follows [22]:

$$r_{j,k} = r_{\max,k} - \frac{j}{k}(r_{\max,k} - r_{\min,k}) \tag{4}$$

with $r_{\max,k}$ and $r_{\min,k}$ are, respectively, the maximum and the minimum radius of the section k. When the orientation angle is equal to 0, the central axis of the unit takes the shape of a circular arc with the center $[d_{j,k}, 0, 0]^T$.

Similarly, the points $A_{i,j,k}$ and $B_{i,j,k}$ where the cables are attached takes a place in an arc of circle of center $[d_{i,j,k}, 0, 0]^T$. Based on these results, we can obtain the following equation:

$$l_{j,k}[mm] = \theta_{j,k} d_{j,k} = \frac{\theta_{j,k}}{\kappa_{j,k}} \tag{5}$$

The different curvature radii of the unit can be described by the Eq. (6) based on **Figure** 7, where $d_{j,k}$ is the radius of curvature of the unit (j, k) and $d_{i,j,k}$ is the radius of curvature for each cable. In addition, there is a relationship between the angle $\varphi_{j,k}$ and $\gamma_{i,k}$ $(i = 1, 2, 3)$.

with $\gamma_{1,k} = -\varphi_{j,k}, \gamma_{2,k} - \frac{2\pi}{3} - \varphi_{j,k}, \gamma_{3,k} = \quad \frac{2\pi}{3} \quad \varphi_{j,k}$. Through the **Figure** 7, the radius of curvature can be expressed as follows:

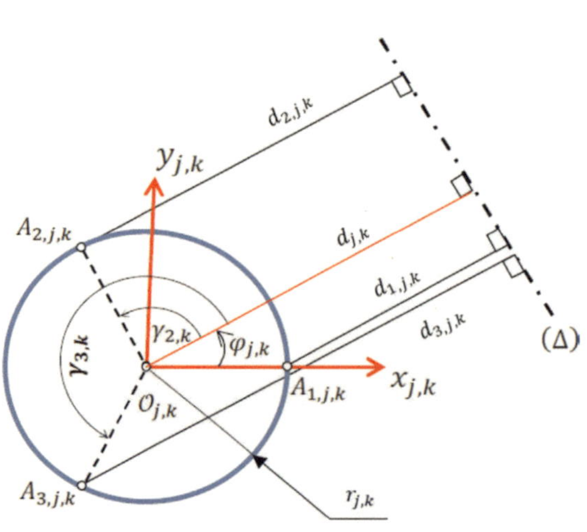

Figure 7.
Definition of the different curvature radii of the unit (j, k).

$$\begin{cases} d_{1,j,k} = d_{j,k} - r_{j,k}\cos\left(\gamma_{1,k}\right) \\ d_{2,j,k} = d_{j,k} - r_{j,k}\cos\left(\gamma_{2,k}\right) \\ d_{3,j,k} = d_{j,k} - r_{j,k}\cos\left(\gamma_{3,k}\right) \end{cases} \tag{6}$$

Based the Eq. (5) and by multiply the three values of the Eq. (6) by $\theta_{j,k}$, we can obtain the following equation:

$$\begin{cases} \ell_{1,j,k} = l_{j,k} - r_{j,k}\theta_{j,k}\cos\left(\gamma_{1,k}\right) \\ \ell_{2,j,k} = l_{j,k} - r_{j,k}\theta_{j,k}\cos\left(\gamma_{2,k}\right) \\ \ell_{3,j,k} = l_{j,k} - r_{j,k}\theta_{j,k}\cos\left(\gamma_{3,k}\right) \end{cases} \tag{7}$$

According to Eq. (7), one way to get $\varphi_{j,k}$ out of this equation is to subtract member by member the second equation from the first equation and the third equation from the first equation as follows:

$$\begin{cases} \ell_{1,j,k} - \ell_{2,j,k} = r_{j,k}\theta_{j,k}\left(\cos\left(\gamma_{2,k}\right) - \cos\left(\gamma_{1,k}\right)\right) \\ \ell_{1,j,k} - \ell_{3,j,k} = r_{j,k}\theta_{j,k}\left(\cos\left(\gamma_{3,k}\right) - \cos\left(\gamma_{1,k}\right)\right) \end{cases} \tag{8}$$

By dividing the equations member by member (8), we obtain the following equation:

$$\frac{\ell_{1,j,k} - \ell_{2,j,k}}{\ell_{1,j,k} - \ell_{3,j,k}} = \frac{\left(\cos\left(\frac{2\pi}{3} - \varphi_{j,k}\right) - \cos\left(-\varphi_{j,k}\right)\right)}{\left(\cos\left(-\frac{2\pi}{3} - \varphi_{j,k}\right) - \cos\left(-\varphi_{j,k}\right)\right)} \tag{9}$$

After applying the trigonometric properties to Eq. (9) we find the following equality

$$\frac{\ell_{1,j,k} - \ell_{2,j,k}}{\ell_{1,j,k} - \ell_{3,j,k}} = \frac{\cos\left(\frac{2\pi}{3}\right)\cos\left(\varphi_{j,k}\right) + \sin\left(\frac{2\pi}{3}\right)\sin\left(\varphi_{j,k}\right) - \cos\left(-\varphi_{j,k}\right)}{\cos\left(-\frac{2\pi}{3}\right)\cos\left(\varphi_{j,k}\right) + \sin\left(-\frac{2\pi}{3}\right)\sin\left(\varphi_{j,k}\right) - \cos\left(-\varphi_{j,k}\right)} \tag{10}$$

After simplification the Eq. (10) becomes:

$$\frac{\ell_{1,j,k} - \ell_{2,j,k}}{\ell_{1,j,k} - \ell_{3,j,k}} = \frac{-\frac{1}{2}\cos\left(\varphi_{j,k}\right) + \frac{\sqrt{3}}{2}\sin\left(\varphi_{j,k}\right) - \cos\left(-\varphi_{j,k}\right)}{-\frac{1}{2}\cos\left(\varphi_{j,k}\right) - \frac{\sqrt{3}}{2}\sin\left(\varphi_{j,k}\right) - \cos\left(-\varphi_{j,k}\right)} \tag{11}$$

In the following, a detailed procedure for obtaining the arc parameters is established.

$1st$ simplification:

$$\begin{aligned} &\frac{1}{2}\ell_{1,j,k}\cos\left(\varphi_{j,k}\right) - \frac{\sqrt{3}}{2}\ell_{1,j,k}\sin\left(\varphi_{j,k}\right) - \ell_{1,j,k}\cos\left(\varphi_{j,k}\right) + \frac{1}{2}\ell_{1,j,k}\cos\left(\varphi_{j,k}\right) \\ &\quad + \frac{\sqrt{3}}{2}\ell_{2,j,k}\sin\left(\varphi_{j,k}\right) + \ell_{2,j,k}\cos\left(\varphi_{j,k}\right) = -\frac{1}{2}\ell_{1,j,k}\cos\left(\varphi_{j,k}\right) + \frac{\sqrt{3}}{2}\ell_{1,j,k}\sin\left(\varphi_{j,k}\right) \\ &\quad - \ell_{1,j,k}\cos\left(\varphi_{j,k}\right) + \frac{1}{2}\ell_{3,j,k}\cos\left(\varphi_{j,k}\right) - \frac{\sqrt{3}}{2}\ell_{3,j,k}\sin\left(\varphi_{j,k}\right) + \ell_{3,j,k}\cos\left(\varphi_{j,k}\right) \end{aligned} \tag{12}$$

2nd simplification:

$$-\sqrt{3}\ell_{1,j,k}\sin\left(\varphi_{j,k}\right)+\frac{1}{2}\ell_{2,j,k}\cos\left(\varphi_{j,k}\right)+\frac{\sqrt{3}}{2}\ell_{2,j,k}\sin\left(\varphi_{j,k}\right)+\ell_{2,j,k}\cos\left(\varphi_{j,k}\right)$$
$$=\frac{1}{2}\ell_{3,j,k}\cos\left(\varphi_{j,k}\right)-\frac{\sqrt{3}}{2}\ell_{3,j,k}\sin\left(\varphi_{j,k}\right)+\ell_{3,j,k}\cos\left(\varphi_{j,k}\right) \tag{13}$$

3rd simplification:

$$-\sqrt{3}\ell_{1,j,k}\sin\left(\varphi_{j,k}\right)+\frac{3}{2}\ell_{2,j,k}\cos\left(\varphi_{j,k}\right)+\frac{\sqrt{3}}{2}\ell_{2,j,k}\sin\left(\varphi_{j,k}\right)$$
$$=\frac{3}{2}\ell_{3,j,k}\cos\left(\varphi_{j,k}\right)-\frac{\sqrt{3}}{2}\ell_{3,j,k}\sin\left(\varphi_{j,k}\right) \tag{14}$$

4th simplification:

$$\frac{\sqrt{3}}{2}\left(-2\ell_{1,j,k}+\ell_{2,j,k}+\ell_{3,j,k}\right)\sin\left(\varphi_{j,k}\right)=\frac{3}{2}\left(\ell_{3,j,k}-\ell_{2,j,k}\right)\cos\left(\varphi_{j,k}\right) \tag{15}$$

5th simplification:

$$\frac{\sin\left(\varphi_{j,k}\right)}{\cos\left(\varphi_{j,k}\right)}=\frac{\sqrt{3}\left(\ell_{3,j,k}-\ell_{2,j,k}\right)}{\left(-2\ell_{1,j,k}+\ell_{2,j,k}+\ell_{3,j,k}\right)} \tag{16}$$

Finally, we obtain following formula the orientation angle:

$$\varphi_{j,k}=\tan^{-1}\left(\frac{\sqrt{3}\left(\ell_{3,j,k}-\ell_{2,j,k}\right)}{2\ell_{1,j,k}-\ell_{2,j,k}-\ell_{3,j,k}}\right) \tag{17}$$

After having determined the orientation angle, we calculate the radius of curvature using the previously mentioned Eq. (7) by summing the three members:

$$3l_{j,k}=\ell_{1,j,k}+\ell_{2,j,k}+\ell_{3,j,k}+r_{j,k}\theta_{j,k}\left(\cos\left(\gamma_{1,k}\right)+\cos\left(\gamma_{2,k}\right)+\cos\left(\gamma_{3,k}\right)\right) \tag{18}$$

Through **Figure** 7, we can write:

$$\sum_{i=1}^{3}\cos\left(\gamma_{i,k}\right)=0 \tag{19}$$

This makes it possible to write:

$$l_{j,k}=\frac{\ell_{1,j,k}+\ell_{2,j,k}+\ell_{3,j,k}}{3} \tag{20}$$

Substituting the Eqs. (5) and (20) into the Eq. (6), and after trigonometric simplifications, the formula of the curvature's radius can be expressed as follows:

$$d_{j,k}=\frac{1}{\kappa_{j,k}}=\frac{r_{j,k}\left(\ell_{1,j,k}+\ell_{2,j,k}+\ell_{3,j,k}\right)}{2\cdot\sqrt{\ell_{1,j,k}^2+\ell_{2,j,k}^2+\ell_{3,j,k}^2-\ell_{1,j,k}\ell_{2,j,k}-\ell_{1,j,k}\ell_{3,j,k}-\ell_{2,j,k}\ell_{3,j,k}}} \tag{21}$$

The last parameter of the arc is the bending angle $\theta_{j,k}$, which can be calculated by substituting the Eqs. (20) and (21) into (5), we then obtain:

$$\theta_{j,k} = \frac{2\cdot\sqrt{\ell_{1,j,k}^2 + \ell_{2,j,k}^2 + \ell_{3,j,k}^2 - \ell_{1,j,k}\ell_{2,j,k} - \ell_{1,j,k}\ell_{3,j,k} - \ell_{2,j,k}\ell_{3,j,k}}}{3r_{j,k}} \tag{22}$$

Finally, the expressions the arc of a circle can be summarized in the Eq. (23):

$$\begin{cases} \varphi_{j,k} = \tan^{-1}\left(\dfrac{\sqrt{3}\left(\ell_{3,j,k} - \ell_{2,j,k}\right)}{2\ell_{1,j,k} - \ell_{2,j,k} - \ell_{3,j,k}}\right) \\ \theta_{j,k} = \dfrac{2\cdot\sqrt{\ell_{1,j,k}^2 + \ell_{2,j,k}^2 + \ell_{3,j,k}^2 - \ell_{1,j,k}\ell_{2,j,k} - \ell_{1,j,k}\ell_{3,j,k} - \ell_{2,j,k}\ell_{3,j,k}}}{3r_{j,k}} \\ \kappa_{j,k} = \dfrac{2\cdot\sqrt{\ell_{1,j,k}^2 + \ell_{2,j,k}^2 + \ell_{3,j,k}^2 - \ell_{1,j,k}\ell_{2,j,k} - \ell_{1,j,k}\ell_{3,j,k} - \ell_{2,j,k}\ell_{3,j,k}}}{r_{j,k}\left(\ell_{1,j,k} + \ell_{2,j,k} + \ell_{3,j,k}\right)} \end{cases} \tag{23}$$

Knowing that or the VC continuum robot each unit has a conical shape, the Eq. (23) must be expressed in terms of the cable lengths $\widehat{\ell}_{i,j,k}$ instead of $\ell_{i,j,k}$. Therefore, according to the assumption mentioned before; each conical shape unit is modeled as an inextensible arc with its individual arc parameters and deforms at small angles, the relationship between these two cable lengths is given by the law of cosines [21]:

$$\widehat{\ell}^{\,2}_{i,j,k} = \ell_{i,j,k}^2 + \left(r_{j-1,k} - r_{j,k}\right)^2 - 2\ell_{i,j,k}\left(r_{j-1,k} - r_{j,k}\right)\cos\left(\beta_{i,j,k}\right) \tag{24}$$

with:

$$\cos\left(\beta_{i,j,k}\right) = \sin\left(\frac{\kappa_{j,k} l_{j,k}}{2}\right)\cos\left(\frac{2}{3}\pi(k-1) - \varphi_{j,k}\right) \tag{25}$$

After solving the Eq. (24), the cables length $\ell_{i,j,k}$ can be expressed as follows:

$$\ell_{i,j,k} = \sqrt{\widehat{\ell}_{i,j,k^2} - \left(r_{j-1,k} - r_{j,k}\right)^2 + \left(r_{j-1,k} - r_{j,k}\right)^2\cos^2\left(\beta_{i,j,k}\right)} + \left(r_{j-1,k} - r_{j,k}\right)\cos\left(\beta_{i,j,k}\right) \tag{26}$$

According to Eq. (26), the length of the cable $\ell_{i,j,k}$ is in function of the cable lenghts $\widehat{\ell}_{i,j,k}$, the disks's radius $\left(r_{j-1,k} - r_{j,k}\right)$ and the angle $\beta_{i,j,k}$. To find an approximate analytical solution between the lengths of the conical unit cable $\widehat{\ell}_{i,j,k}$ and the configuration state $\kappa_{j,k}$, the influence of $\beta_{i,j,k}$ on the lengths of the cylindrical unit cable according to Eq. (26) must be negligible. Since the unit's conical shape cannot be changed, angles $\beta_{i,j,k}$ close to $\frac{\pi}{2}$ are required. This can be achieved by choosing a high number of units, because the bending angle $\theta_{j,k}$ of the unit decreases with an increasing number of units per section. In this case, the Eq. (26) simplifies to:

$$\ell_{i,j,k} = \sqrt{\hat{\ell}_{i,j,k^2} - \left(r_{j-1,k} - r_{j,k}\right)^2} \tag{27}$$

2.3.3 Independent transformation

To express the cartesian coordinates $\left(x_{j,k}, y_{j,k}, z_{j,k}\right)$ in terms of the arc parameters, we substitute Eq. (23) into (3), the cartesian coordinates $\left(x_{j,k}, y_{j,k}, z_{j,k}\right)$ as a function of the cable lengths can be expressed as follows:

$$\begin{cases} x_{j,k} = \dfrac{r_{j,k}\left(\ell_{1,j,k} + \ell_{2,j,k} + \ell_{3,j,k}\right)}{M_{j,k}} \sin^2\left(\dfrac{M_{j,k}}{3r_{j,k}}\right) \cos\left(\tan^{-1}\left(\dfrac{\sqrt{3}\left(\ell_{3,j,k} - \ell_{2,j,k}\right)}{2\ell_{1,j,k} - \ell_{2,j,k} - \ell_{3,j,k}}\right)\right) \\ y_{j,k} = \dfrac{r_{j,k}\left(\ell_{1,j,k} + \ell_{2,j,k} + \ell_{3,j,k}\right)}{M_{j,k}} \sin^2\left(\dfrac{M_{j,k}}{3r_{j,k}}\right) \sin\left(\tan^{-1}\left(\dfrac{\sqrt{3}\left(\ell_{3,j,k} - \ell_{2,j,k}\right)}{2\ell_{1,j,k} - \ell_{2,j,k} - \ell_{3,j,k}}\right)\right) \\ z_{j,k} = \dfrac{r_{j,k}\left(\ell_{1,j,k} + \ell_{2,j,k} + \ell_{3,j,k}\right)}{2M_{j,k}} \sin\left(\dfrac{2M_{j,k}}{3r_{j,k}}\right) \end{cases} \tag{28}$$

with: $M_{j,k} = 2 \cdot \sqrt{\ell_{1,j,k}^2 + \ell_{2,j,k}^2 + \ell_{3,j,k}^2 - \ell_{1,j,k}\ell_{2,j,k} - \ell_{1,j,k}\ell_{3,j,k} - \ell_{2,j,k}\ell_{3,j,k}}.$

2.3.4 FKM of a single section

As mentioned before, a section k is a concatenation of m units. Moreover, each section deforms without torsional effect, so we can have the forward kinematic model of a section by the multiplication of transformation matrices, which can be expressed by the following relation:

$$T_k^{k-1} = T_{m,k}^{0,k} = \prod_{j=1}^{m} T_{j,k}^{j-1,k} \tag{29}$$

where $T_{j,k}^{j-1,k}$ is the (4 × 4) matrix defining the orientation and position of the origin of the reference frame $R_{j,k}$ in the frame $R_{j-1,k}$.

For a VC continuum robot, an approximate formula is developed in [22], which allows to express the bending angle of the robot's units in function of the first bending angle of the first unit in the same section. The developed formula is expressed as follows [22]:

$$\theta_{j,k} = \frac{r_{1,k}}{r_{j,k}}\theta_{1,k} \tag{30}$$

As the number of units increases, the accuracy of the solutions of the Eq. (30) also increases, that is, when the angle β_i is closer to $\frac{\pi}{2}$ (**Figure 3**). Therefore, using Eq. (30), the forward kinematic model of a single section can only be expressed by two variables $\theta_{1,k}$ and φ_k. This approximation leads to a remarkable simplification and reduction of the number of variables involved in the FKM.

2.3.5 FKM of the multi-section conical robot

Based on the Eq. (30), the identification of the robot's end-effector posture becomes relatively simple because the different unit angles will be expressed in terms of the bending angle of the first unit. Thus the forward kinematic model for a multi-section continuum robot (see **Figure 1**) can be easily found by successive multiplication of the independent transformation matrices for each section k and the transformation matrix of the static frame of reference. The resulting matrix is given by the following equation:

$$T_n^{orig} = \begin{bmatrix} A_n^{orig} & O_n^{orig} \\ 0_{1\times3} & 1 \end{bmatrix} = T_0^{orig} \prod_{k=1}^{n} T_k^{k-1} \tag{31}$$

where the matrices T_0^{orig} and T_k^{k-1} represent the static transformation matrix and the independent transformation matrix of each flexible section respectively.

3. Workspace generation for a continuum robot

By definition, the workspace is the set of all positions accessible by the robot's end effector.

Figure 8 shows the possible positions that can be reached by the first section of the continuum robot, as well as by the continuum robot with two sections in two-dimensional representations. These points were obtained using the FKM, by varying the first bending angle for each section in the interval $\theta_{1,k} \in \left[-\frac{\pi}{8}, \frac{\pi}{8}\right]$.

On **Figure 8**, we present some possible positions that the continuum robot's end effector can reach, which are designated by the letter A (dotted curved lines in red) as

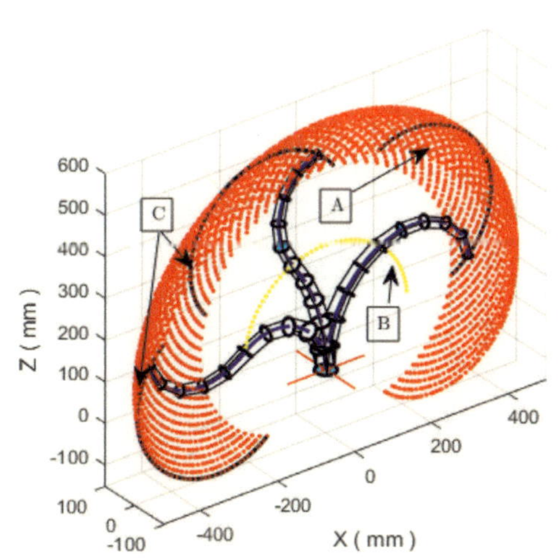

Figure 8.
2D view of the continuum robot workspace at VC.

well as the positions of the first section of the robot designated by the letter B (dotted curved lines in yellow).

3.1 Comparison between the workspace of a continuum robot with VC and CC

In this section, we compare the workspace of a flexible continuum robot with constant and variable curvature. Lets start with the comparison of a single section of the continuum robot, the characteristics of each section are shown on 1. The first bending angle and orientation of the section are varied in the interval $\left[0, \frac{\pi}{5}\right]$ and $[-\pi, \pi]$ respectively. As it can be seen in **Figures 9** and **10**, the idea of variable

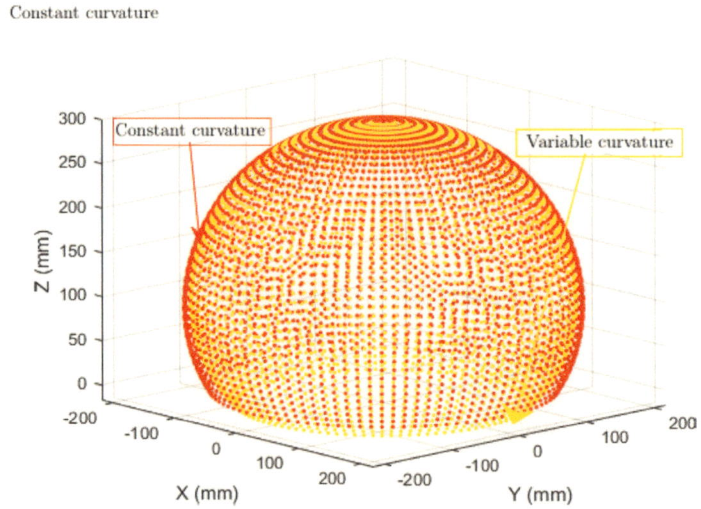

Figure 9.
3D view of the workspace of a CC and VC section.

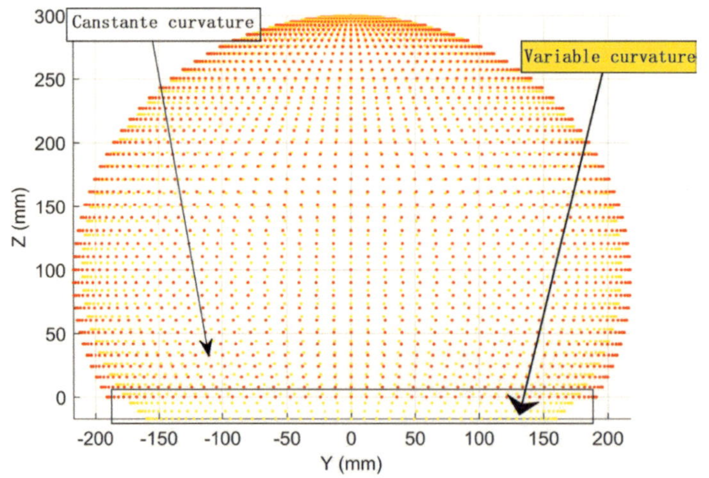

Figure 10.
View in the YZ plane of the workspace of a CC and VC section.

curvature is able to produce bending angles with increasing curvature, namely, the end point of the section can reach much more curved positions with respect to the constant curvature.

In the second example, we vary the first bending angle for each section in the same range $\theta_{j,k} \in \left[-\frac{\pi}{8}, \frac{\pi}{8}\right]$.

Figures 11 and **12** show the workspace of a flexible continuum robot with CC and VC respectively. Comparatively, the end effector of a VC flexible continuum robot can reach much more curved positions than a CC continuum robot as shown in **Figure 13**.

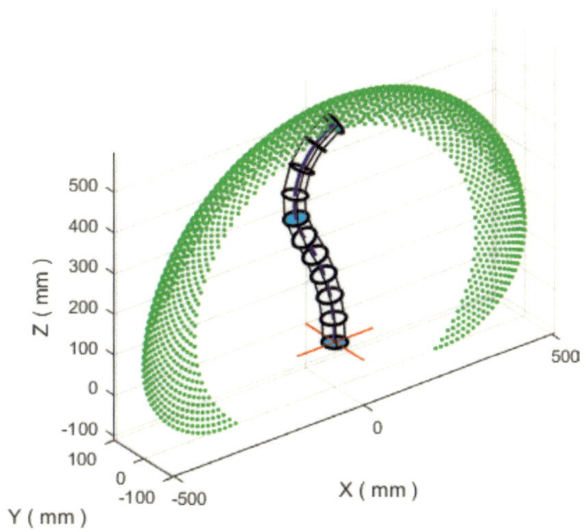

Figure 11.
2D workspace view for a flexible continuum robot with constant curvature: $r_{\max} = r_{\min} = 25mm$.

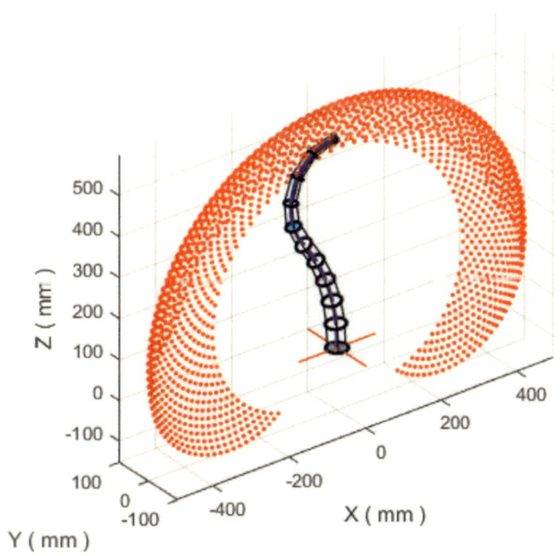

Figure 12.
2D workspace view for a flexible continuum robot with variable curvature: $r_{\max} = 25mm, r_{\min} = 10mm$.

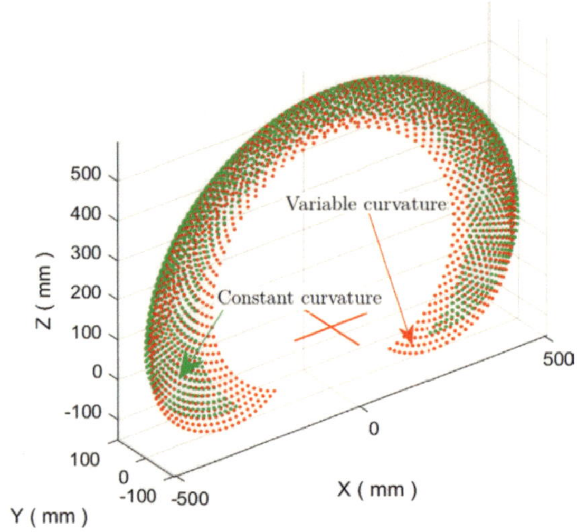

Figure 13.
Comparison of workspace for a flexible continuum robot with CC and VC.

4. Inverse kinematic modeling

For any robotic system, the Inverse Kinematic Model IKM can be summarized as follows: for a given posture (position and orientation), we determine the necessary actuation variables that satisfy certain movement constraints. In the continuum robots' world, the inverse kinematic models are a very nonlinear problem, which is inferred by calculating the inverse function of forward kinematic models. IKMs can easily orient robots by cartesian position equations rather than curvature and rotational equations. But obtaining IKMs remains a difficult task, especially when it comes to multi-section continuum robots. In almost cases IKMs cannot be analytically calculated and have several problems that lie in the following:

- Unavailability of solutions when the target points are outside the robot's workspace.

- Several solutions or endless solutions when the target point can be reached by different configurations.

For the robots studied in this chapter, the inverse kinematic model consists in calculating the cable lengths and geometric parameters corresponding to the desired position of the robot's end-effector. To this end, the artificial neural networks and meta-heuristic optimization will be used to solve the inverse kinematic model of a single section continuum robot and a multi-section continuum robot respectively. The development of the used approaches to solve the IKMs is done through the forward kinematic models.

4.1 ANN for solving the IKM of single section continuum robots with VC

Neural Networks have achieved a great success in many areas due to their learning and generalization capabilities as well as their parallelism. They have been successfully used in many applications, such as classification, noise filtering, system modeling and control ... etc. One of the fields where NN has received an increasing interest is that of solving the robots' IKMs.

Due to the complexity of continuum robots models, very few works which aim at solving their IKMs have been carried out, especially with varible curvature. To this end in this section, the Multi-Layer Perceptron (MLP) neural networks will be used and developed to find the IKMs of single section continuum robots with variable curvature. The hidden layers contains neurons with sigmoid activation function and a linear activation function in output layers. For the sake of avoiding the complexity of the neural networks a minimum number of neurons and hidden layers that gives the good learning are chosen.

The forward kinematic model is calculated for the sake of obtaining the P_{x_i}, P_{y_i}, and P_{z_i} coordinates of the robot's end tip which themselves are used to train the proposed MLP. the used objective function for training neural networks is given by the following Eq. (32):

$$F = \frac{1}{N}\sum_{i=1}^{N}\left((P_{x_i} - X_{c_i})^2 + \left(P_{y_i} - Y_{c_i}\right)^2 + (P_{z_i} - Z_{c_i})^2\right) \tag{32}$$

Where N represent the number of data. X_{c_i}, Y_{c_i}, and Z_{c_i} represent the spatial coordinates of a located position on the prescribed trajectory. P_{x_i}, P_{y_i}, and P_{z_i} represent the position of the robot's end tip for each specific position of the prescribed trajectory, which are obtained from the FKM. Explicitly, they represent the three first components of the forth column of Eq. (31) which describe the FKM of the considered robots.

4.1.1 Outline and simulation

Two simulations examples are considered, for the first simulation; a single section continuum robot with variable curvature must follow up arc-shaped trajectory and for the second simulation the robot must follow up circular trajectories.

The characteristics of the considered robot are given in **Table 1**.

4.1.2 IKM of single section continuum robot

The FKM of a single section is first calculated using Eq. (31). After that, to find the IKM of the considered robot the orientation angle is calculated analytically using

Parameter	value	Description
m_k	5 units	Number of units per section
L_k	300 mm	Total length of the section
$r_{\min,k}$	17.5 mm	Radial cable distance
$r_{\max,k}$	25 mm	Radial cable distance

Table 1.
Parameters of the single section flexible continuum robot with VC.

the FKM by dividing the position P_y by the position P_x. Finally, the static MLP (**Figure 14**) with the following configuration is used to calculate the bending angles:

- The input layer contains four inputs (x, y, z, ϕ)
- One hidden layer with 20 neurons.
- One linear output neuron that gives the approximated θ.

Learning rate: $\alpha = 0.01$.

- Type of optimizer: Levenberg-Marquardt backpropagation.

A random dataset that covers all-inclusive possible positions which can be attained by the robot's first section, namely its workspace, are generated using the FKM to train the developed model.

The training result error value to the bending angle is given by **Figure 15**. To evaluate the effectiveness of the obtained IKM. A second randomly dataset that cover all-inclusive possible positions in the robot's workspace is generated using the FKM. **Figure 16** gives the error test results of the developed neural model for the bending angle θ. **Figures 15** and **16** show the accuracy and the effectiveness of the obtained IKM. The values of the mean square error (MSE) of the achieved model are given in **Figure 17**.

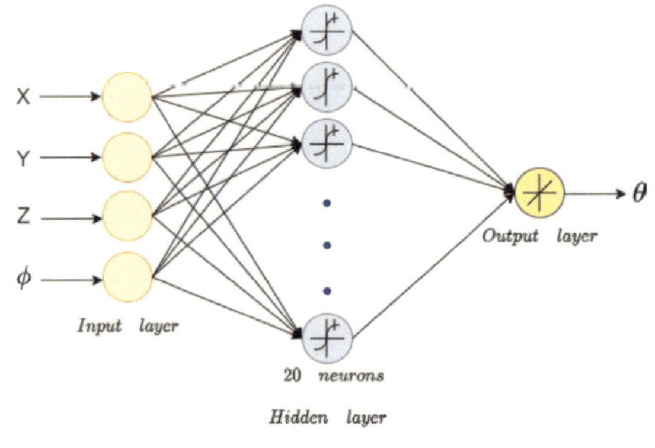

Figure 14.
Neural network structure for spacial single section continuum robot.

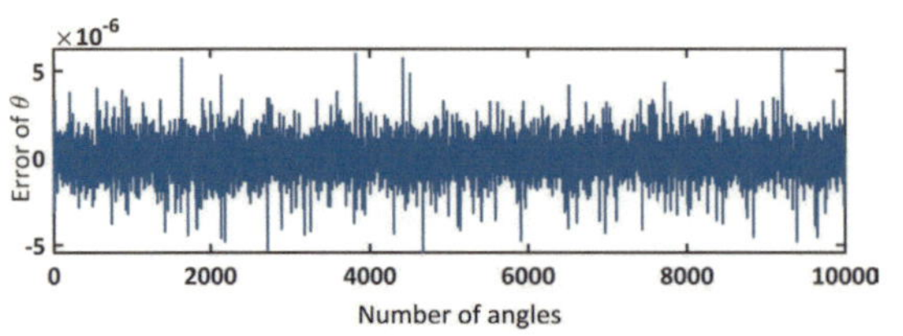

Figure 15.
Training error results of the neural model for the bending angle θ.

After tested the effectiveness of the obtained model (IKM), the considered robot is tested on practically realistic-like trajectories (circular and arc-shaped trajectories) in order to assess the robot's behavior via 3D simulations.

As it is shown in **Figure 18**, the considered single section continuum robot with VC can accurately follow the arc-shaped trajectory within its 3D workspace (yellow curved arcs).

To show the accuracy of the generated trajectory, euclidean errors are calculated. As it is shown in **Figure 19**, the maximum error between the generated and the

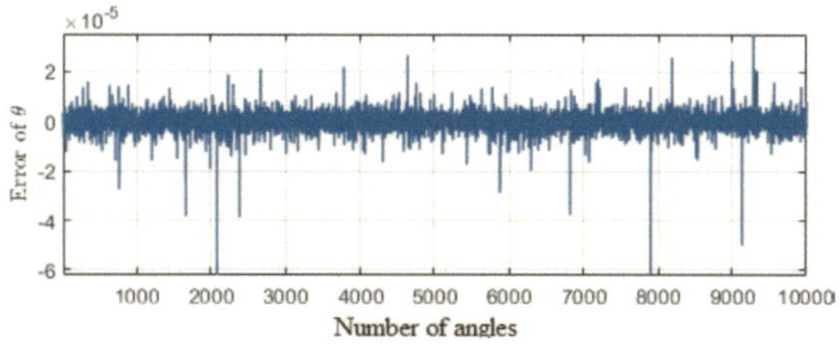

Figure 16.
Error test results of the neural model for the bending angle θ.

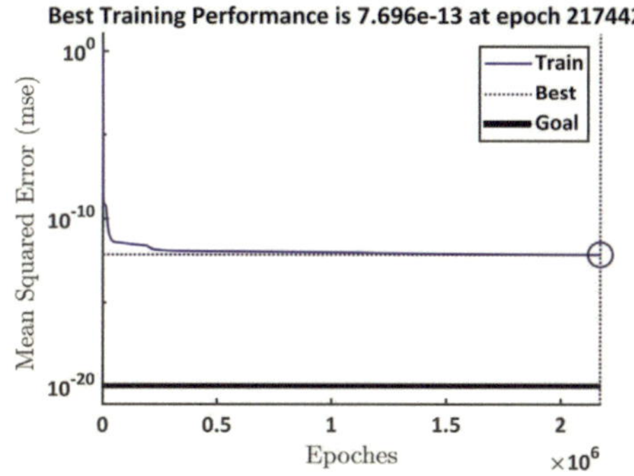

Figure 17.
Training performance of the model.

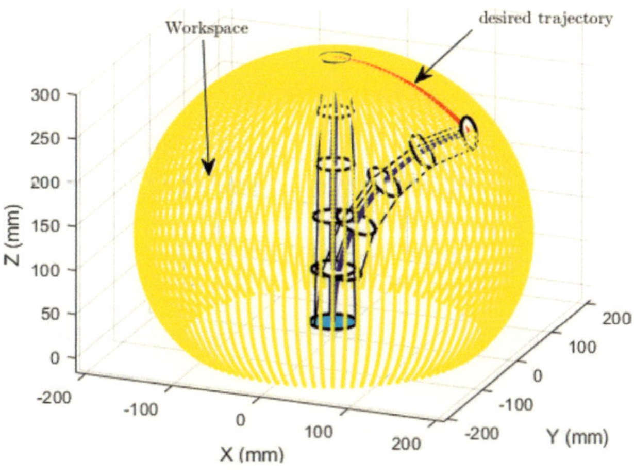

Figure 18.
Two configurations of a single section continuum robot following an arc shaped trajectory within its 3D workspace.

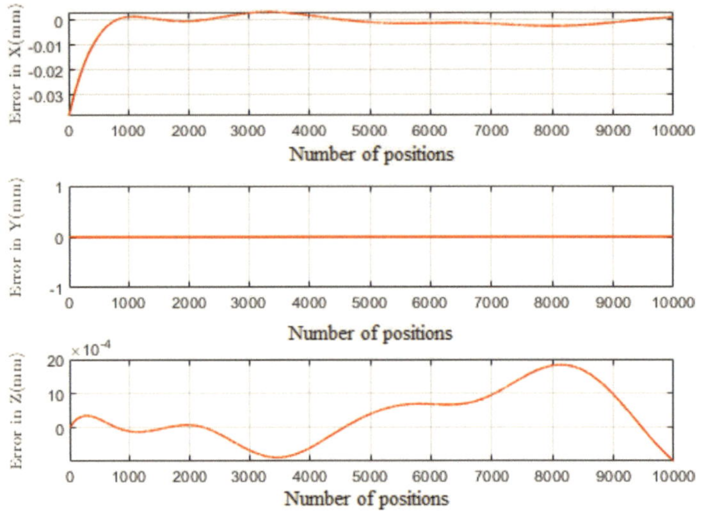

Figure 19.
Error between the reference trajectory (arc-shaped) and the trajectory (arc-shaped) given by the NN.

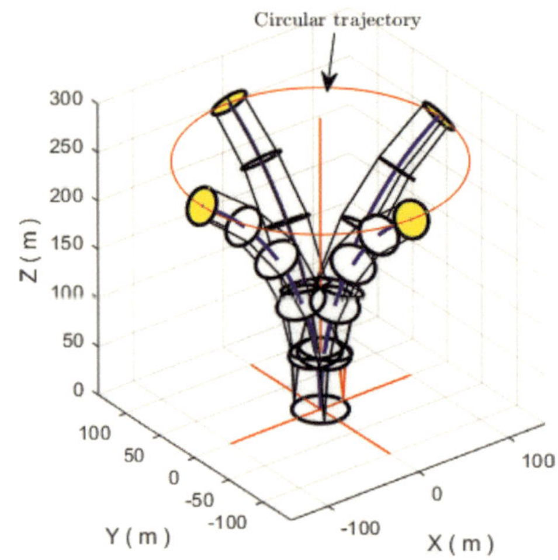

Figure 20.
Representation of the continuum robot's first section tracking the desired circular trajectory.

desired trajectory is of the order of 0.002 mm (**Figure 20**). Similarly to the first simulation example, the continuum robot track a circular trajectory within its workspace (See **Figure 20**). **Figure 21**, shows the euclidean error of the trained NN.

4.2 Methaheuristic optimization for solving the IKM of multi section continuum robots with VC

The main objective of adopting meta-heuristic optimization techniques unlike artificial neural networks to solve the inverse kinematic model of multi-section continuum robots is the ability to deal with the complexity of these robot's models and their endless solutions, especially when including orientation angles. As well as their

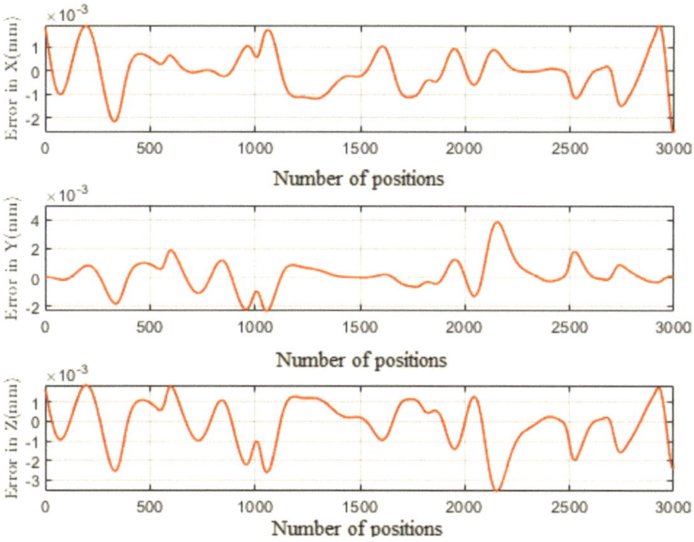

Figure 21.
Error between the reference circular trajectory and the circular trajectory given by the NN.

capabilities to make the robot avoid obstacles simply by taking the appropriate configuration to accomplish the required task. To this end in this section, a meta-heuristic optimization algorithm called teaching learning-based optimization (TLBO) is used to solve the inverse kinematic model of a multi-section continuum robot with the ability to make the robot dexterously avoid obstacles.

4.2.1 Teaching learning based optimization algorithm

TLBO Algorithm is originally inspired from the process' effect of teaching a set of students. It was proposed by Rao et al. [27] and Rao and Savsani [28]. The algorithm consists of two phases, teacher and learners phases. In the process of this algorithm, a set of learners is considered as population, the whole population is provided with various topics, which are supposed to be the variables of the optimization problem. The learners' results illustrate the fitness value of the optimization problem. The outstanding solution among the whole population is considered to be the teacher, while the variables are the objective function's parameters and its best value is the best solution to the considered problem.

4.2.2 Teacher phase

In the first phase, the teacher thoroughly tries to improve the mean result of the whole class according to the taught subject. At each attempt (iteration) i, there are m topics and n_p learners in a specific topic j. The best learner in the classroom is chosen as the teacher. The difference $d^{\mathrm{i}}_{\mathrm{j,k}}$ between the existing mean results of each topic and the teacher result for each subject is given by:

$$d^{\mathrm{i}}_{\mathrm{j,k}} = r\left(X^{\mathrm{i}}_{\mathrm{j,k_{best}}} - T_F M^{\mathrm{i}}_{\mathrm{j}}\right) \tag{33}$$

where

$$M_{\mathrm{j}}^{\mathrm{i}} = \frac{\sum_{\mathrm{k}=1}^{n} X_{\mathrm{j,k}}^{\mathrm{i}}}{n} \tag{34}$$

and $X_{\mathrm{j,k_{best}}}^{\mathrm{i}}$ is the best selected learner's result in a topic j, T_F is a factor which can be either 1 or 2, r is a random number in the range [0,1], $M_{\mathrm{j}}^{\mathrm{i}}$ is the mean result. Based on Eq. (33), the existing solution is updated in the teacher phase as follows:

$$Xnew_{\mathrm{j,k}}^{\mathrm{i}} = X_{\mathrm{j,k}}^{\mathrm{i}} + d_{\mathrm{j,k}}^{\mathrm{i}} \tag{35}$$

where $Xnew_{\mathrm{j,k}}^{\mathrm{i}}$ is the updated value of $X_{\mathrm{j,k}}^{\mathrm{i}}$ which is sent to the learner phase.

4.2.3 Learner phase

In this particular phase, the learners enhance their level of knowledge by a random interaction among themselves. The learner can improve his/her knowledge by interacting with a learner who has more knowledge than him/her. The whole process can be mathematically formulated as follows:

$$\begin{cases} Xnew_{\mathrm{j},A}^{\mathrm{i}} = X_{\mathrm{j},A}^{\mathrm{i}} + r\left(X_{\mathrm{j},A}^{\mathrm{i}} - X_{\mathrm{j},B}^{\mathrm{i}}\right) & if \quad F_A^{\mathrm{i}} < F_B^{\mathrm{i}} \\ Xnew_{\mathrm{j},A}^{\mathrm{i}} = X_{\mathrm{j},A}^{\mathrm{i}} + r\left(X_{\mathrm{j},B}^{\mathrm{i}} - X_{\mathrm{j},A}^{\mathrm{i}}\right) & if \quad F_B^{\mathrm{i}} < F_A^{\mathrm{i}} \end{cases} \tag{36}$$

where A and B are two randomly selected learners, F_A^i and F_B^i are the fitness values of X_A and X_B for A and B, respectively. It is noteworthy to say that the aforementioned Eqs. ((29), (30), (33)) are valid exclusively when it comes to minimization problems.

4.2.4 Comparison of TLBO with other optimization algorithms

Generally, almost all optimization methods require specific parameters that have a crucial role on the algorithm's functionality as well as its performance. **Table 2** shows the required parameters for some metaheuristic algorithms. Unlike other optimization techniques, TLBO does not require any specific parameters, which makes it a handy algorithm to deal with the complicated problems of engineering. TLBO uses the best

Algorithms	Required parameters
GA	Crossover probability, mutation rate, selection method
PSO	Acceleration constants, weight variation, maximum velocity
ABC	Onlooker bees number, employed bees number, food sources number
HS	Range of each variable, pich rate, number of improvisation

Table 2.
The required parameters for each algorithm.

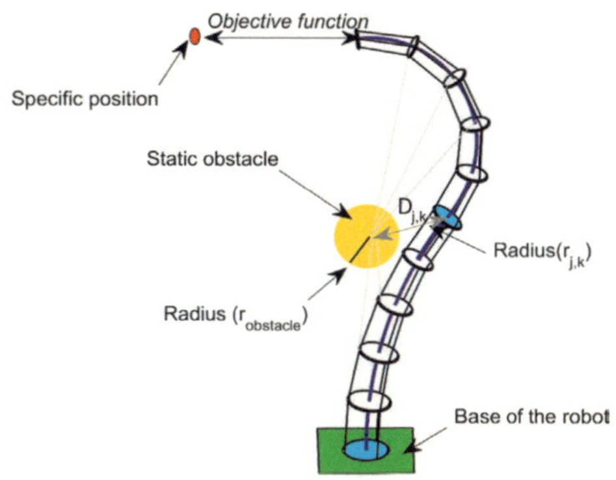

Figure 22.
Global view of the governing objective function and the adopted strategy for the obstacle avoidance.

solution at each iteration to shift to the best existing solution in the population, which paves the way to a fast convergence.

4.2.5 Objective function and problem formulation

The resolution of the IKM can be addressed using an objective function, which describes the distance between the prescribed position and the robot's end tip (**Figure 22**), where the lowest possible distance can be considered as a solution for a specifically prescribed position. The cost function contains the desired position coordinate and the robot's end tip coordinates it can be mathematically expressed as follows:

$$F = \sqrt{(P_{X_i} - X_{c_i})^2 + (P_{Y_i} - Y_{c_i})^2 + (P_{Z_i} - Z_{c_i})^2} \tag{37}$$

Furthermore, X_{c_i}, Y_{c_i}, and Z_{c_i} represent the spatial coordinates of a located position on the prescribed trajectory. P_{X_i}, P_{Y_i}, and P_{Z_i} represent the position of the robot's end tip obtained from the FKM. These positions represent the three first components of the fourth column of the matrix (31).

$$D_{\mathrm{j,k}}^2 \leq \left(r_{obstacle} + r_{\mathrm{j,k}}\right)^2 \tag{38}$$

Where $r_{obstacle}$ is the radius of the spherical-like shape obstacle, and $r_{j,k}$ is the radius of the selected robot's disks. The obstacle avoidance can be fulfilled if the distance between the robot's disks radius summed up with the obstacle's radius is bigger than $D_{j,k}$. The studied robot in this section is given by 3., which is consists of two sections $n = 2$. Each section has five disks with different diameters; therefore, the robot's 3D workspace is the zone where the operation of searching for a specific solution is performed by TLBO (**Table 3**).

4.2.6 TLBO implementation to solve the IKM of VL continuum robot

Three Simulation examples through Matlab are carried out. For the former, a 3D simulation of a two-section continuum robot with variable length following two types

	Section (k = 1)	Section (k = 2)	Description
m_k	5 units	5 units	Number of units per section
L_k	300 mm	300 mm	Total length of the section
$r_{min,k}$	17.5 mm	10 mm	Radial cable distance
$r_{max,k}$	25 mm	17.5 mm	Radial cable distance

Table 3.
Parameters of the two-section flexible continuum robot.

of trajectories, namely a linear and spherical trajectory, is considered. The needed variables to be found by the TLBO are (θ_1, φ_1) for the robot's first section and (θ_2, φ_2) for the robot's second section. The second simulation addresses a free-collision example during the follow up of a circular trajectory. In the third simulation, a 3D simulation of a three-section continuum robot during the follow up of a linear trajectory is carried out.

4.2.7 Step of the optimization algorithm

- Step 0: initialization

    - Choose the population size (n_p), the number of variables (m), the termination criterion (ϵ), the maximum number of iteration (k_{max}), the admissible maximum and minimum values for each variable $\left(X_{\mathrm{j}_{max}}, X_{\mathrm{j}_{min}}\right)$.

    - For each point of the reference trajectory do the following steps:

- Step 1: initial solution

For $\mathrm{k} = 1 : n_p$.
For $\mathrm{j} = 1 : m$

- Choose randomly the initial solution as follows: $x_{\mathrm{j,k}}^1 = rand(0, 1) * \left(X_{\mathrm{j}_{max}} - X_{\mathrm{j}_{min}}\right) + X_{\mathrm{j}_{min}}$ where: r is a random number between 0 and 1

    * End

        - End

        - Select the desired reference trajectory point.

Set $\mathrm{i} = 1$

- Step 2: teacher phase

For $\mathrm{k} = 1 : n_p$

* Evaluate the corresponding objective function (F_{k}) using Eq. (37).

- End

- Select the student who gives the minimum value of the objective function $\left(F_{k_{best}}\right)$ as a teacher.

For j = 1 : m

- Calculate the mean result $\left(M_j^i\right)$ using Eq. (34).

- End

For j = 1 : m

- Calculate the difference mean $\left(d_j^i\right)$ using Eq. (33).

- End

For k = 1 : n_p.
For j = 1 : m

- Calculate the new solutions $\left(Xnew_{j,k}^i\right)$) using Eq. (35).

* End

- End

For k = 1 : n_p

* Evaluate the new corresponding objective function $(Fnew_k)$ using the new solutions $\left(Xnew_{j,k}^i\right)$ according to Eq. (37).

- End

- Apply a greedy selection between the new and the old solutions.

- Step 3: learners phase

- Choose randomly a q pairs of solutions.

For k = 1 : n_p.
For j = 1 : m

- Calculate the new solutions $\left(Xnew_{j,k}^i\right)$ using Eq. (36)

* End

  - End

For k = 1 : n_p

- Evaluate the new corresponding objective function $(Fnew_k)$ using the new solutions $\left(Xnew_{j,k}^{i}\right)$ according to Eq. (37)

  - End

  - Apply a greedy selection between the new and the old solutions.

- Step 4: iterative process

  - If $F_{k_{best}} < \epsilon$ or $i > k_{max}$

- Go to step 5

  - Else

$i = i + 1$

- Go to step 2

  - End if

- Step 5: obtaining the final solution

  - Apply the computed best solution.

  - Go to step 1.

4.2.8 Simulation of a two-section continuum robot following a circular trajectory in the presence of an obstacle

In this application, a two-section continuum robot with variable curvature is considered to follow a circular trajectory (see **Figure 23**), which is defined by Eq. (39), in the presence of a static obstacle as previously explained in **Figure 22**. The robot has to follow the prescribed circular trajectory in the condition that it does not collide or smoothly touch the spherical trajectory.

$$\begin{cases} X = 30\,cos\left(\frac{\pi}{5}t\right) \\ Y = 30\,sin\left(\frac{\pi}{5}t\right) \\ Z = 440 + 30\,sin\left(\frac{\pi t}{5}\right) \end{cases} \tag{39}$$

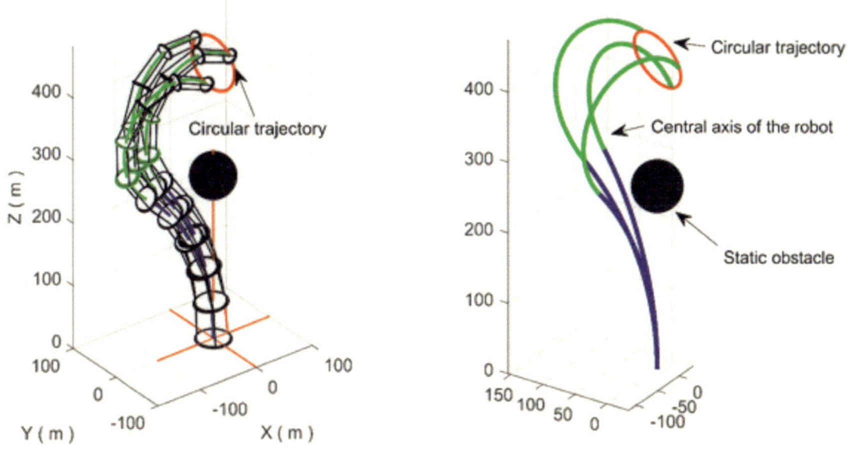

Figure 23.
(left) VL continuum robot following a circular trajectory in the presence of a static obstacle; (right) different configuration of the central axis of the robot.

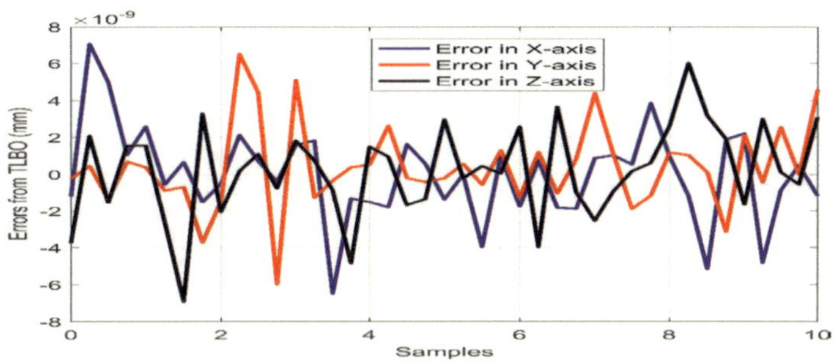

Figure 24.
Error between the desired circular trajectory and the generated one by TLBO.

with t = 0:0.25:10 As it can be seen from **Figure 23**, the variable curvature continuum robot follows the circular trajectory with a free collision. The whole operation is performed based on TLBO algorithm which generates the required angles allowing the robot to properly follow the circular trajectory with respect to the length variation. Remarkably, based on the Euclidean error between the generated trajectory by TLBO and the desired trajectory (**Figure 24**), TLBO can be considered as a powerful optimization algorithm for solving the inverse kinematic model of continuum robots.

Figure 25 represents the cables length allowing the robot to follow the circular trajectory with free collision.

It is noteworthy to say that the calculation of the cables length is mainly based on angles which are calculated from the proposed algorithm and on the strategy of calculating the parallel robots' IKMs, in other words, the three cables for each unit are considered as the three segments connecting the parallel robot's platforms, namely two platforms are linked through three segments with prismatic articulation.

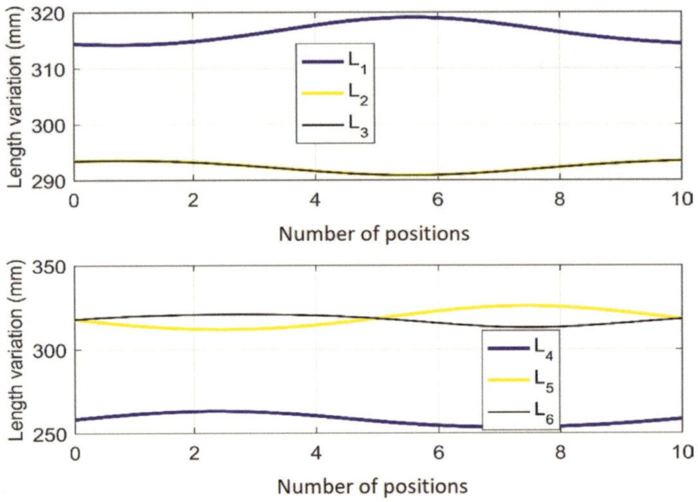

Figure 25.
The cables length generated by TLBO ensuring the free-collision during circular trajectory.

5. Conclusion

The aim of this chapter was to develop methods for modeling of continuum robots, which can provide satisfactory and effective performance. In fact, neural networks and optimization methods have been widely used for modeling and controlling nonlinear systems. The secret of neural networks' success lies in the fact that any nonlinear system can be modeled with a certain precision using a simple neural network having learning and generalization capabilities. On the other hand, the emergence of many optimization methods has made it possible to consider more complex optimization problems for which numerical methods cannot give acceptable solutions.

In this chapter, the different types of continuum robots and their modeling were properly studied. This study allowed us to examine the limitations and advantages of the adopted approaches and maked it easier for us to suggest several ways to improve the modeling and control methods based on artificial intelligence tools. The accomplished work can be summarized as follows:

- A recently developed meta-heuristic approach so-called teaching learning based optimization was adopted to solve the inverse kinematic model of a continuum robot with variable curvature. The TLBO algorithm is not fussy and does not require any specific parameters to be tuned up. In other words, unlike the other developed meta-heuristic algorithms which need specific parameters to perform correctly, TLBO can be considered as the best optimization method to tackle the complexity of continuum robots. To emphasize, based on the available state of art about optimization methods, it is concluded that the TLBO algorithm is way better accurate than the other optimization algorithms for solving the IKM.

- An artificial neural network is developed to solve the inverse kinematic model of single-section continuum robot with variable curvature. The development of the used artificial neural network is done through the forward kinematic model which is mainly used to train the ANN. At the training of the neural network is performed based on a random generated database from the FKM. The obtained neural network is tested on solving the IKM of continuum robots through various examples. The bending angles which allow the robot to the trained properly follow a desired trajectory can be easily and accurately generated by ANN.

Acknowledgements

This work is supported by the DGRSDT (Direction Générale de la Recherche Scientifique et du Développement Technologique), Algeria.

Parts of this chapter were previously published in the doctoral thesis by Abdelhamid GHOUL: Contribution to the modeling and control of continuum and deformable robots using artificial intelligence tools. 2022. UNIVERSITÉ DE BLIDA 1, Faculté de Technologie, Département d'Automatique et Électrotechnique. Available from: https://di.univ-blida.dz/jspui/handle/123456789/25562

Conflict of interest

The authors declare no conflict of interest.

Author details

Abdelhamid Ghoul[1*†], Selman Djeffal[2†] and Kamel Kara[1]

1 Saad Dahleb University, Blida, Algeria

2 National Polytechnic School of Constantine, Constantine, Algeria

*Address all correspondence to: abdelhamid_ghoul@yahoofr

† These authors contributed equally.

References

[1] Wilkening P, Alambeigi F, Murphy RJ, Taylor RH, Armand M. Development and experimental evaluation of concurrent control of a robotic arm and continuum manipulator for osteolytic lesion treatment. IEEE Robotics and Automation Letters. 2017;**2**(3):1625-1631

[2] Mahoney A, W, Anderson PL, Swaney PJ, Maldonado F, Webster RJ. Reconfigurable parallel continuum robots for incisionless surgery. In: 2016 IEEE/RSJ International Conference on Intelligent Robots and Systems (IROS). Daejeon, Korea (South): IEEE; 2016. pp. 4330-4336

[3] Tonapi M, M, Godage IS, Walker ID. Next generation rope-like robot for in-space inspection. In: 2014 IEEE Aerospace Conference. Big Sky, MT, USA: IEEE; 2014. pp. 1-13

[4] Bajo A, Simaan N. Finding lost wrenches: Using continuum robots for contact detection and estimation of contact location. In: 2010 IEEE International Conference on Robotics and Automation. Anchorage, AK, USA: IEEE; 2010. pp. 3666-3673

[5] Orekhov AL, Black CB, Till J, Chung S, Caleb D, Rucker. Analysis and validation of a teleoperated surgical parallel continuum manipulator. IEEE Robotics and Automation Letters. 2016; **1**(2):828-835

[6] Farshid Alambeigi Y, Wang SS, Gao C, Murphy RJ, Iordachita I, Taylor RH, et al. A curved-drilling approach in core decompression of the femoral head osteonecrosis using a continuum manipulator. IEEE Robotics and Automation Letters. 2017;**2**(3):1480-1487

[7] Tingyu Q, Chen J, Shen S, Xiao Z, Yue Z, Lau HYK. Motion control of a bio-inspired wire-driven multi-backbone continuum minimally invasive surgical manipulator. In: 2016 IEEE International Conference on Robotics and Biomimetics (ROBIO). Qingdao, China: IEEE; 2016. pp. 1989-1995

[8] Fraś J, Czarnowski J, Maciaś M, Główka J, Cianchetti M, Menciassi A. New STIFF-FLOP module construction idea for improved actuation and sensing. In: 2015 IEEE International Conference on Robotics and Automation (ICRA). Seatle, WA, USA: IEEE; 2015. pp. 2901-2906

[9] Cianchetti M, Ranzani T, Gerboni G, De Falco I, Laschi C, Menciassi A. STIFF-FLOP surgical manipulator: Mechanical design and experimental characterization of the single module. In: 2013 IEEE/RSJ International Conference on Intelligent Robots and Systems. Tokyo, Japan: IEEE; 2013. pp. 3576-3581

[10] Jessica Burgner-Kahrs D, Rucker C, Choset H. Continuum robots for medical applications: A survey. IEEE Transactions on Robotics. 2015;**31**(6): 1261-1280

[11] Conrad B, L, Jung J, Penning RS, Zinn MR. Interleaved continuum-rigid manipulation: An augmented approach for robotic minimally-invasive flexible catheter-based procedures. In: 2013 IEEE International Conference on Robotics and Automation. Karlsruhe, Germany: IEEE; 2013. pp. 718-724

[12] Li J, Teng Z, Xiao J. Can a continuum manipulator fetch an object in an unknown cluttered space? IEEE Robotics and Automation Letters. 2016;**2**(1):2-9

[13] Webster III RJ, Jones BA. Design and kinematic modeling of constant

curvature continuum robots: A review. The International Journal of Robotics Research. 2010;**29**(13):1661-1683

[14] Jones BA, Walker ID. Kinematics for multisection continuum robots. IEEE Transactions on Robotics. 2006;**22**(1): 43-55

[15] Iqbal S, Mohammed S, Amirat Y. A guaranteed approach for kinematic analysis of continuum robot based catheter. In: 2009 IEEE International Conference on Robotics and Biomimetics (ROBIO). Guilin, China: IEEE; 2009. pp. 1573-1578

[16] Gravagne I, A, Walker ID. Kinematic transformations for remotely-actuated planar continuum robots. In: Proceedings 2000 ICRA. Millennium Conference. IEEE International Conference on Robotics and Automation. Symposia Proceedings (Cat No. 00CH37065). Vol. 1. San Francisco, CA, USA: IEEE; 2000. pp. 19-26

[17] Abdelhamid Ghoul, Kamel Kara, Mohamed Benrabah, and Boualem Nasri. Control of Continuum Robot Using Two Optimized PID Controllers. In 2021 Multi Conference on Electrical Engineering CEE. 2021

[18] Ghoul A, Kara K, Benrabah M, Hadjili ML. Optimized nonlinear sliding mode control of a continuum robot manipulator. Journal of Control, Automation and Electrical Systems. 2022;**33**(5):1355-1363

[19] Ghoul A, Kara K, Djeffal S, Benrabah M, Hadjili ML. Inverse kinematic model of continuum robots using artificial neural network. In: 2022 19th International Multi-Conference on Systems, Signals and Devices (SSD), setif, Algeria. 2022. pp. 1893-1898

[20] Ghoul A, Djeffal S, Kara K, Aouaichia A. Dynamic modeling and control of continuum robots using an optimized PID control. In: 2023 International Conference on Advances in Electronics, Control and Communication Systems (ICAECCS), Blida, Algeria. 2023. pp. 1-6

[21] Mahl T, Hildebrandt A, Sawodny O. A variable curvature continuum kinematics for kinematic control of the bionic handling assistant. IEEE Transactions on Robotics. 2014;**30**(4): 935-949

[22] Djeffal S, Amouri A, Mahfoudi C. Kinematics modeling and simulation analysis of variable curvature kinematics continuum robots

[23] Ghoul A, Kara K, Djeffal S, Benrabah M, Hadjili ML. Artificial neural network for solving the inverse kinematic model of a spatial and planar variable curvature continuum robot. Archive of Mechanical Engineering. 2022;**69**(4):595-613

[24] Djeffal S, Ghoul A, Morakchi MR, Mahfoudi C, Belkedari M. Optimized computer torque control and dynamic model of a spatial single section continuum robot. Results in Control and Optimization. 2023;**12**: 100264

[25] Djeffal S, Mahfoudi C, Amouri A. Comparison of three meta-heuristic algorithms for solving inverse kinematics problems of variable curvature continuum robots. In: 2021 European Conference on Mobile Robots (ECMR). IEEE. 2021. pp. 1-6

[26] Escande C. Towards modeling of a class of bionic manipulator robots. 2013

[27] Venkata Rao R, Savsani VJ, Vakharia DP. Teaching–learning-based optimization: A novel method for

constrained mechanical design optimization problems. Computer-Aided Design. 2011;**43**(3):303-315

[28] Venkata Rao R, Savsani VJ, Vakharia DP. Teaching–learning-based optimization: An optimization method for continuous non-linear large scale problems. Information Sciences. 2012; **183**(1):1-15

Chapter 6

Fostering Creativity via AI Robotics

Um Albaneen Jamali

Abstract

This research paper presents a research study exploring the impact of robotics programs on fostering creativity in primary school pupils. The study conducted over 4 months and employed a mixed methodology combining pre-post CAP testing, interviews, and observational analysis. The study sample comprised 60 female students, aged 10 to 12, from a middle socio-economic background. The participants were randomly divided into control and treatment groups. The students in the treatment group were introduced to an Arduino robotics program involving artificial intelligence, whereas the participants in the control group did not join such program. The findings from the pre-posttests demonstrated that the AI robotics program, significantly enhanced thinking skills of creativity including flexibility, fluency, elaboration, and originality. The paper offers insights for education policy makers and provides recommendations for future research in this field. The chapter recommends incorporating AI robotics in the curriculum while calling educational policy makers for proving training and research opportunities in this field.

Keywords: AI, robotics, creativity, elementary school, education, Arduino, flexibility, fluency, elaboration originality

1. Introduction

The quest for long-term sustainability has triggered a global search for innovative solutions to resource limitations, a task that is inherently connected to creativity. As the pulse of sustainable development, creativity is not only about imagination and ingenuity but extends to incorporating new technologies and innovative ways of using existing ones. The fourth industrial revolution has paved the way for significant creativity and innovation, necessitating a redesign of educational programs to meet the evolving demands. This paper first explores the concept of creativity and its associated skills and then describes educational programs designed to foster creativity with a focus on robotics and artificial intelligence (AI). The paper examines a research study investigating the effect of AI on enhancing creative thinking skills among primary school girls and offers insights for future research.

IntechOpen

2. Creativity

Creativity involves the generation of something that is useful, original, and novel [1]. While the concept has various interpretations and models, these variations do not imply contradiction or confusion but rather showcase the different ways creativity has been explored and elaborated in diverse contexts [1].

2.1 Creative thinking skills

Creative thinking is associated with divergent, convergent, and emergent thinking patterns and skills [2, 3]. Divergent thinking generates original ideas. Convergent thinking assesses the novelty and usefulness of these ideas, and emergent thinking translates these novel and useful ideas into creative outputs. These thinking patterns necessitate various creative thinking skills: fluency (generating multiple ideas), flexibility (producing diverse thoughts), originality (creating novel ideas), and elaboration (adding details to the final product) [4].

3. Developing creativity in education

Numerous educational programs have been developed to nurture creativity in education. These programs used two main approaches. Teachers either directly introduce creative thinking skills independently from the curriculum or indirectly integrate them into curriculum-related activities. However, some of these programs are outdated, being either too lengthy, de-motivating, lacking innovation, or failing to connect to modern technological advancements. On the other hand, AI programs showed increasing attention by educators [5]. Such programs have been mostly linked to STEM (science, technology, engineering, and mathematics) subjects and used the main methodology of project-based learning PBL. This is largely because this methodology embraces the teacher introducing a problem and encouraging students to examine, analyze, and reflect to find the creative solution. In this perspective, each student is the protagonist in developing their own reflections and innovative conclusions from experimentation in a critical and creative manner [6]. Similarly, the current research used Arduino in STEM projects using the methodology of PBL. It investigated the impact of AI to foster creativity. The paper shall now explore AI robotics and related research in the field.

3.1 AI educational robotics

Robots integrated with AI offer innovative solutions to challenges across various fields, including education. The AI robots, which were used for the purpose of the current study, are equipped with diverse environmental sensors such as proximity and soil moisture sensors besides vision devices that provide real-time data for analysis and action. Arduino includes a programmable circuit board and software for writing and uploading simplified C++ code to the board. The AI capability of Arduino robots allows them to collect and analyze information, make inferences based on their environment and mission, and act accordingly to deliver optimal results. AI-powered robots are more efficient, saving time and energy guaranteeing accuracy, and minimal mistakes [7].

4. Studies on Arduino robotics and creativity

Arduino robotics and creativity have been a focal point in numerous research studies over the past decade. A systematic literature review on prototyping with Arduino indicated its benefits to teachers, students, and amateur innovators ranging from building and programming to creativity and innovation [8]. Arduino has a simple, inexpensive, and easy-to-understand open-source platform for individuals with interest on interactive hardware and software projects. Some research studies on Arduino robotics and creativity in education shall now be highlighted in some detail.

Tiryaki and Adigüzel [9] examined the effects of STEM-based robotics applications on the creativity and attitude of students toward science. In their mixed-methods study, 60, and seventh-grade students, in two groups of experimental (n = 30) and control groups (n = 30) attended a post-school course, which consisted of 2 weeks of pre-applications and 4 weeks of applications. The "Torrance Creative Thinking Test" and TOSRA were applied to measure creativity and attitude toward science, respectively, as pre- and posttest. Furthermore, data from semi-structured interviews were analyzed using content analysis. The results indicated significant gains in students' creativity and attitudes toward science due to their participation in STEM-based robotics applications. They argued that "STEM-based robotic applications are more effective in increasing students' creativity than the inquiry-based learning model" (p. 295). Although they applied Torrance Creativity Thinking Test and noted that the test measures creative thinking skills of fluency, flexibility, detailing or elaboration, and originality; they did not attend to each thinking skills and rather indicated the overall creativity scores of their participants.

Handayani and Setya [10] investigated the impact of Arduino for learning STEM on creativity of preservice physics teachers in laboratory of design physics. Teacher participants were first taught the basics of Arduino and its coding, four mini-projects, and theoretical explanations about sensors and data generating process. A qualitative method with the narrative design was used for the purpose of this study. Data was collected *via* observation. A sample of 14 preservice physics teachers (5 males and 10 females) participated in the study. The training course was conducted virtually due to COVID-19 pandemic. Teachers were asked to design a laboratory activity to support learning physics using sensor and Arduino. The tool used to collect data was creativity product analysis matrix (CPAM) measuring three creativity dimensions in terms of elaboration, resolution, and novelty dimension. Teachers' performance on design creativity of the physics laboratory demonstrated development in three dimensions of problem-solving and resolution, elaboration, and originality.

Jamali [11] in her longitudinal study investigated the impact of AI intervention comparing to LEGO robotic intervention on developing creative thinking skills. She conducted her study using a mixed method of pre-posttests and observation among a sample of 60 female students aged 10 to 12, who were randomly selected into two groups of treatment and controlled groups. The treatment group participated in a LEGO robotics program in the first year and an Arduino robotics program in the second year. The control group did not participate in any robotics program. The results indicated that the AI robotics program enhanced all the creativity thinking skills, including fluency, flexibility, originality, and elaboration. LEGO robotic intervention fostered fluency, flexibility, and elaboration, while failed to statistically and significantly develop originality. Jamali recommended using AI in education and called for future research in this field.

Guven et al. [12] sought to understand the impact of performing coding activities using Arduino-assisted robotics on students' attitude toward robotics, creativity in science, and motivation. Their research adopted a mixed method, using a sample of 11 and sixth graders (6 females and 5 males) who participated in a STEM program. Quantitative data was collected using scientific creativity scales, motivation scales, and robotics attitude scales, while qualitative data was gathered through semi-structured interviews. The results showed an increase in students' creativity, motivation and attitude toward robotics, and motivation. Furthermore, students exhibited thinking skills of creativity such as, flexibility, fluency, elaboration and originality, and demonstrated problem-solving abilities in real-life situations The researchers Guven and her colleagues recommended the use of Arduino-assisted robotics coding applications in the teaching of sixth-grade science curricula for sixth-grade students.

Kim and Lee [13] investigated the effects of incorporating Arduino into an educational intervention on promoting creativity. A total of 20 high school students (n = 20) participated in the intervention over a span of 36 hours. However, the findings demonstrated no significant gain in the students' creative problem-solving abilities. Students partcipants' feedback following the intervention revealed that according to them, the Arduino-based program although was intriguing and somewhat satisfying, but felt overwhelming by the challenges in designing and debugging. The researchers suggested that teaching materials, pedagogical methods, and activities need to be reconsidered before incorporating AI into education.

Chou [14] examined students' learning performances in an Arduino-based educational robotics program. A learning environment was set up in a public elementary school in Taiwan, where 30 fifth-grade students took part in an after-school club for a period of 16 weeks. The students were grouped randomly into control and experimental groups. Students in the experimental group joined a weekly AI robotics educational course, while those in the control group and participated in others, such as homework. Both observation and pre-posttest designs were used to evaluate the participants' coding and problem-solving skills, and electrical engineering knowledge. The quantitative results suggested that AI robotics courses significantly developed the participants' problem-solving skills while enhancing their knowledge in coding and electrical engineering. Qualitative results from observation showed that students used a variety of alternative approaches (i.e. flexibility) and unique thoughts (i.e., originality) in building AI robots to resolve real-life challenges. The researcher highlighted the comfort and ease of using Arduino solutions in daily life and underscored the significance of supporting students in both hardware and software debugging considering the Arduino's limitations in this regard.

However, the aforementioned studies had certain limitations. Some were conducted in controlled settings with children participating in an intense AI robotics program a short time-span. Some studies, such as Guven et al. relied on using applications for coding activities while did not attend to engendering parts of building robots. Moreover, most research relied heavily on quantitative methods, while the samples were not representative (particularly in terms of gender) and were small in numbers, . In addition, some research did not investigate the development of skills of creative thinking in terms of flexibility, fluency, elaboration, and originality using standardized tests, while others mentioned fostering creative thinking skills as secondary data. Utilizing standardized creativity tests would offer a basis for comparing results from past, present, and future studies. To address these weaknesses, the current research examined changes in students' creativity before and after implementing robotics interventions in their classroom over 12 weeks, with a sample of 60 female students. The research hypothesized that AI robotics interventions would positively impact the development of creative thinking skills of flexibility, fluency, elaboration, and originality among students.

5. Methods

5.1 Research design and sample

The current study adopted a mixed-methods approach. The qualitative component incorporated observation and interviews, with the researcher observing students as they designed and coded robots. Additionally, 30-minute semi-structured interviews were conducted, during which students were asked to provide further details about their robots. The quantitative component employed a control treatment and pre-posttest design.

The study's sample consisted of 60 female students, aged between 9 and 12 years (n = 60). These students were from a local all-girls primary school in Bani Jamra, originating from middle socio-economic backgrounds. The participants were randomly divided into control and treatment groups. Prior to the study, participants' creativity levels were assessed using the CAP [15] test.

5.2 Procedures

The program was delivered in weekly two-hour sessions over the course of 12 weeks. The intervention consisted of an Arduino robotics program, utilizing artificial intelligence (AI) techniques. It was delivered in a similar manner as the first year's program but lasted 10 weeks. For both years, the control group did not receive any intervention. The Frank Williams [13] Creativity Assessment Package (CAP) was used to measure students' creativity before and after the AI interventions. Throughout the sessions, the researcher observed students' performances and conducted 30-minute semi-structured interviews, where students provided more details about their robots.

5.3 Data analysis

Triangulation was ensured through mixed methods. The research adopted a pre-posttest designing. Quantitative data was gathered from participants' performances on CAP tests before and after the AI robotics interventions. The CAP test consisted of 12 frames, each containing a line as a thought provoker, and children were asked to draw up to 12 drawings using these lines. Creative thinking skills such as flexibility, fluency, elaboration, and originality were evaluated through these drawings. The resulting data was analyzed using a repeated measures t-test. Qualitative data incorporated observations of how children designed and coded robots, emphasizing the creative thinking skills of fluency, flexibility, elaboration and originality. Transcripts from participants' interviews informed the observations.

6. Results

The study yielded interesting results from qualitative and quantitative data. The students in the treatment group who participated in the AI robotics interventions demonstrated significantly improved creative thinking skills compared to the control group. The results shall now be discussed in more detail.

The AI robotics intervention also showed promising results. Paired samples t-tests revealed significant improvements in posttest creativity scores for the treatment group compared to the control group, $t(29) = 2.630$, $p = .014$ ($p < .05$). As shown in **Table 1**, the treatment group's scores ($M = 34.300$, $SD = 11.744$) increased by an average of 7.4 points compared to the control group ($M = 26.900$, $SD = 8.095$). This indicates that AI robotics programs can enhance overall creativity, corroborating previous research (**Table 1**) [7–12, 14].

Moreover, significant increases were observed in scores regarding developing creative thinking skills of fluency (i.e., developing many ideas) in the treatment group at posttest, $t(29) = 2.630$, $p = .008$ ($p < .05$). Evidence from students' performances and interview transcripts supported the findings. For example, a group of students described how they designed an AI bed, which performed many tasks, such as alarming, washing the face, feeding, riding to the school and so on and so forth. This demonstrated that AI robotics programs can foster the development of fluency supporting earlier mentioned research studies [9, 12, 14].

The t-test also revealed significant gains in flexibility scores at posttest, with p values of .035. The treatment group exhibited significant improvements in flexibility, $t\ (29) = 2.217$, $p = .035$ ($p < .05$) compared to the control group. Examples of fostering flexibility were evident in students AI projects. For instance, when they asked to develop AI tools for teaching real-life Math examples; a group of students used soil moisture sensor and an LCD screen, which was designed and programmed to indicate high (above 100%) and low (below 50%) soil moisture if the soil was wet or dry respectively. The results supported the hypothesis that AI robotics interventions can enhance flexibility supporting the findings of previous studies [11, 12, 14]. This suggested that AI robotics programs can develop flexibility.

In addition, significant gains were indicated in elaboration scores in the treatment group at posttest, $t(29) = 2.305$, $p = .029$ ($p < .05$). Evidence from students' performances and interview transcripts supported the findings. Examples of student projects such as a Smart Cap that featured 18 components and performed four activities using AI, indicated the development of creative thinking skills of elaboration. This demonstrated that AI robotics programs can foster the development of elaboration, supporting the findings of previous studies [10–12, 14].

Furthermore, students displayed a significant increase in originality after the AI intervention, $t(29) = 2.189$, $p = .037$ ($p < .05$). Qualitative evidence also backed this advancement in originality. For example, treatment group students designed a baby car seat, which analyzed baby's crying and respond accordingly. When the crying was low (what students called boredom cry) AI was playing a baby song and when the crying was loud (what students called fear cry) AI was entertaining the baby by moving the toys. This unique idea, emerging from elementary-level students, was showcased in a national techno exhibition for secondary level and university students. These findings were aligning with research that supports the role of AI robotics intervention in fostering originality and unique idea generation. Students attributed this growth in originality to the freedom of choosing from a range of sensors. As one student put it, "I feel Arduino robots are real! You can ask for many demands because there are plenty sensors one can choose from, as long as you open your imagination." These results suggest that AI robotics programs foster originality, supporting previous studies [10–12, 14] but contradicting study [13].

Paired Samples Test									
		Paired Differences							Significance
					95% Confidence Interval of the Difference				
		Mean	Std. Deviation	Std. Error Mean	Lower	Upper	t	df	Two-Sided p
Pair 1	OverallCreativity - OverallCreativitypost	−7.40000	15.41070	2.81360	−13.15445	−1.64555	−2.630	29	.014
Pair 2	Fluency - Fluencypost	−1.96667	3.78275	.69063	−3.37917	−.55416	−2.848	29	.008
Pair 3	Flexibility - Flexibilitypost	−1.23333	3.04770	.55643	−2.37137	−.09530	−2.217	29	.035
Pair 4	Originality - Originalitypost	−.66667	1.66782	.30450	−1.28944	−.04389	−2.189	29	.037
Pair 5	Elaboration - Elaborationpost	−3.76667	8.95076	1.63418	−7.10894	−.42440	−2.305	29	.029

Table 1.
The results of the t-test with repeated measures comparing students' posttest scores after AI robotics intervention on creative thinking skills in the treatment and controlled group.

7. Discussion

The results of this study offer insights into the influence of AI robotics on student creativity development. Posttest scores from the treatment group, who received AI robotics interventions, showed significant enhancements in creative thinking skills compared to those from the control group, who received no AI intervention. Qualitative data from task observation and student interviews also strengthened the study's findings, as treatment group students frequently discussed their enhancements in fluency, flexibility, originality, and elaboration. These findings suggest that AI robotics interventions positively impacted students' creative abilities, echoing previous research [2, 7–12, 14] that found similar effects.

When the intervention program incorporated AI robotics, significant improvements were recorded across all creativity skills — fluency, flexibility, originality, and elaboration. This aligns with earlier studies [10, 12] that underscored the positive impact of AI robotics interventions on creativity. The benefits of AI, such as speed and accuracy, deep learning capabilities, and the ability to handle more complex tasks, might explain this. Equipped with various sensors (e.g., proximity, humidity, sound sensors, accelerometer, and other environmental sensors), AI-powered robots can sense, analyze, and respond to data in real-time, making them seem "real."

The present study was more successful at developing a broader range of creative thinking skills compared to previous research. This could be due to the small sample size, the duration of the AI robotics intervention, and the setting where the research was conducted. The research was carried out in the comfort of the participants' classroom. The instructions were delivered by their teacher, unlike some research studies where unfamiliar coaches -provided instructions in unfamiliar afternoon club settings. Given the students' age, a familiar setting and coach might be beneficial for fostering creativity.

The study spanned 12 weeks, allowing students to master AI programming without undue pressure. On the other hand, the research had its weaknesses. The sample included only female students from a local primary school, which served children from predominantly middle socio-economic background. Performing further research with diverse samples including both genders, various academic level students, and with diverse socio-economic statuses could yield comprehensive insights into the effects of AI robotics in future research.

The present research applied both qualitative and quantitative methods. Employing other means of data collection, such as case studies, may offer additional info and deeper our understanding on the way AI robotics enhance children's creativity.

8. Conclusions

In conclusion, this paper underscores the influence of AI robotics in fostering creative thinking skills. The research offered valuable insights into the effect of AI robotics on the development of creative thinking skills. The AI robotics intervention suggested a broad impact on creative thinking skills, including fluency, flexibility, originality, and elaboration. AI was found to be beneficial in providing a realistic environment for creativity development. Consequently, policy makers and educators may consider several implications. For instance, policy makers might consider offering training in AI robotics as part of professional development programs or funding large-scale and longitudinal studies in the field. School leaders and teachers might consider integrating AI robotics into their curricula. It is worth noting that teaching AI robotics requires well-trained instructors to prevent student demotivation due to bugs in AI. Similarly, manufactures may consider resolving the bugs with Arduino hardware and provide easy-to-use and applicable AI hardware for educational purposes. This study demonstrated attempts at enhancing sustainable development through fostering creativity. More efforts are needed to incorporate AI robotics into education, thereby fostering creativity and enhancing sustainable development in the near future.

Conflict of interest

The authors declare no conflict of interest.

Notes/thanks/other declarations

Special Thanks to Zeeba Khayamy and my beloved students, my princess of creativity.

Author details

Um Albaneen Jamali[1,2,3,4]

1 Education of Gifted and Talented Children, The Kingdom of Bahrain

2 World Council for Gifted and Talented Children (WCGTC), The Kingdom of Bahrain

3 The Arabian Council for Gifted and Talented Children, The Kingdom of Bahrain

4 Educational Management and Leadership, The University of Nottingham, England

*Address all correspondence to: 29jamali@gmail.com

References

[1] Williams R, Runco MA, Berlow E. Mapping the themes, impact, and cohesion of creativity research over the last 25 years. Creativity Research Journal. 2016;**28**(4):385-394. DOI: 10.1080/10400419.2016.1230358

[2] Kim KH, Coxon S. Fostering creativity using robotics among students in STEM fields to reverse the creativity crisis. In: Demetrikopoulos MK, Pecore JL, editors. Interplay of Creativity and Giftedness in Science. Rotterdam, The Netherlands: Sense Publishers; 2016. pp. 351-365. DOI: 10.1007/978-94-6300-163-2_19

[3] Kozlowski JS, Chamberlin SA. Raising the bar for mathematically gifted students through creativity-based mathematics instruction. Gifted and Talented International. 2019;**34**(1-2):79-90. DOI: 10.1080/15332276.2019.1690954

[4] Paek SH, Sumners SE. The indirect effect of teachers' creative mindsets on teaching creativity. Journal of Creative Behavior. 2019;**53**:298-311. DOI: 10.1002/jocb.180

[5] Fidai A, Capraro MM, Capraro RM. "Scratch"-ing computational thinking with Arduino: A meta-analysis. Thinking Skills and Creativity. 2020;**38**:100726. DOI: 10.1016/j.tsc.2020.100726

[6] García-Tudela PA, Marín-Marín J-A. Use of Arduino in primary education: A systematic review. Education Sciences. 2023;**13**(2):134. DOI: 10.3390/educsci13020134

[7] Chhaya K, Khanzode A, Sarode RD. Advantages and disadvantages of artificial intelligence and machine learning: A literature review. International Journal of Library & Information Science. 2020;**9**(1):30-36

[8] Kondaveeti HK, Kumaravelu NK, Vanambathina SD, Mathe SE, Vappangi S. A systematic literature review on prototyping with Arduino: Applications, challenges, advantages, and limitations. Computer Science Review. 2021;**40**:100364. DOI: 10.1016/j.cosrev.2021.100364

[9] Tiryaki A, Adigüzel S. The effect of STEM-based robotic applications on the creativity and attitude of students. Journal of Science Learning. 2021;**4**(3):288-297

[10] Handayani W, Setya W. Preservice physics teachers' laboratory project design: Arduino-based creativity. AIP Conference Proceedings. 2022;**2468**(1):020012. DOI: 10.1063/5.0102773

[11] Jamali UY. Developing creativity via LEGO & AI robotics. In: Wyk M, editor. Technology in Learning. London, UK, London: Intechopen; 2023. DOI: 10.5772/intechopen.110277

[12] Guven G, Cakir NK, Sulun Y, Cetin G, Guven E. Arduino-assisted robotics coding applications integrated into the 5E learning model in science teaching. Journal of Research on Technology in Education. 2022;**54**(1):108-126. DOI: 10.1080/15391523.2020.1812136

[13] Kim SW, Lee Y. Development and application of Arduino-based education program for high school students. Journal of Theoretical & Applied Information Technology. 2017;**95**(18):4367-4375

[14] Chou P-N. Skill development and knowledge acquisition cultivated by maker education: Evidence from Arduino-based educational robotics.

Eurasia Journal of Mathematics, Science and Technology Education. 2018;**14**(10):em1600. DOI: 10.29333/ejmste/93483

[15] Williams FE. Creativity Assessment Packet Examiner's Manual. Austin, TX: PRO-ED; 1993. p. 13

Section 3

Robotics in Clinic and Surgery

Chapter 7

Clinical Significance of Robot Manipulators for Grasp, Balance, and Gait Recovery (from the Medical Point of View)

Ivet B. Koleva, Radoslav R. Yoshinov and Borislav R. Yoshinov

Abstract

Robots and virtual reality are now introduced in clinical neurorehabilitation practice. The inclusion of robots in the complex rehabilitation program supports functional recovery, ameliorates the quality of care, and increases the quality of life of patients. Grasp, balance, and gait are important activities, necessary for autonomy in everyday life. Therefore, grasp and gait recovery are important indices of the efficacy of neurorehabilitation. The application of robotic rehabilitation accelerates the process of grasp and gait training, improves the balance and gait stability, and ameliorates the independence of patients in everyday life. We have clinical experience with patients suffering from different neurological and neurosurgical conditions: hemiparesis after cerebral vascular accident, multiple sclerosis, and Guillain-Barre syndrome and after neurosurgical intervention for brain or spinal cord aneurysm, tumor, or trauma. Here, we present some of our own results in patients with post-stroke hemiparesis and with inferior paraparesis after spinal cord injury. We consider that the introduction of robots in the process of grasp, balance, and gait recovery of patients with neurological conditions and after neurosurgical interventions can be very useful for neurorehabilitation clinical practice.

Keywords: neurorehabilitation, robotics, sensors, grasp, balance, gait, autonomy, activities of daily living, hemiparesis, paraparesis

1. Introduction

Robots and virtual reality are now introduced in clinical neurorehabilitation practice. We consider that the inclusion of robots in the complex rehabilitation program accelerates functional recovery, ameliorates the quality of care, and increases the quality of life of patients.

2. Rehabilitation, physical and rehabilitation medicine, and neurorehabilitation

2.1 Rehabilitation

Rehabilitation is a functional therapy, based on detailed functional assessment. The goals of rehabilitation are oriented to functional compensation and functional recovery, according to the World Disability Report of the World Health Organization and the World Bank [1]. Currently, in clinical rehabilitation practice, we apply the International Classification of Diseases (ICD) [2], the International Classification of Functioning, disability and Health (ICF) [3], and appropriate scales for the evaluation of the quality of life [4].

2.2 Physical and rehabilitation medicine

The medical specialty of Physical Medicine (or Physiatry) uses different physical modalities for the improvement of the patient's quality of life. The principal objectives of Physical and Rehabilitation Medicine (PRM) are oriented to the "promotion of physical and cognitive functioning," to "amelioration of autonomy and quality of life" and to "optimization of social participation" of patients with health-related disability and co-morbidity, according to the documents of the PRM-section of the European Union of Medical Specialists [5–8]. For this, we need the support of rehabilitation medicine, rehabilitation therapy, and assistive technologies [1].

2.3 Neurorehabilitation algorithms

Neurorehabilitation (NR) is an interdiscipline – between neurology, neurosurgery, physical and rehabilitation medicine [9, 10].

The NR algorithm includes a detailed functional assessment and a complex program of care (with different natural and pre-formed physical factors) [4].

The functional assessment of the patient includes evaluation of the range of motion (goniometry), of muscle force and muscle weakness (dynamometry and manual muscle test); assessment of grasp, balance, gait, and autonomy in everyday life; ICD diagnostics and ICF evaluation.

The NR program is established by a synergic combination of several physical modalities: physical exercises, sports, activities, cryo- and thermo-procedures, mineral waters and peloids, electric currents, light (including laser), magnetic field, ultrasound, etc. [4, 9, 10]. In clinical practice, usually we prescribe *one or two procedures with pre-formed modalities; one thermo- or cryo-agent; one or two physio-therapeutic methods* (soft tissue techniques, manual therapy, analytic exercises); and *one or two ergo therapeutic activities* (sports; training of grasp, balance, gait or activities of daily living; self-care; domestic occupations; and professional activities) [9].

At the end of the NR course, a final functional assessment is required – with the goal to evaluate the efficacy of our intervention and to prescribe consecutive rehabilitation procedures. In chronic conditions, usually, we recommend an everyday program of physical exercises, combined with a number of rehabilitation courses of 10–30 procedures (for a period of 2 weeks to 1 month), applied periodically.

3. Grasp and gait: grasp and gait rehabilitation

3.1 Grasp, grip and prehension

Grasp is the ability of the digits to seize, for holding or picking up different objects, using the hand. The firm grasp is defined as *grip*. *Prehension* is the act of grasping or seizing; the name descends from the Latin verb "prehendere," meaning "to seize" or "to grasp". The traditional classification of Hippocrates is based on function. Hippocrates differentiates *power grasp or grip* for strength; and *prehension grip or pinch* for precision [10–12].

In rehabilitation clinical practice, we apply the conventional classification of Neumann, based on the purpose of the task and the number of involved fingers. He differentiates power and precision grasp; grip, pinch, and hook grip. All fingers are involved in grip; thumb and index are used in pinch; all fingers without thumb participate in hook grip. Investigators and practitioners divide *power grasp* into: cylindrical, spherical, and hook grips. In a cylindrical grasp, the entire palmar surface of the hand is positioned around a cylindrical shaped object, using many muscles: finger flexors, including thumb flexors, intrinsic muscles, and thumb abductors. During spherical grasp, the thenar and hypothenar eminences are in a cupping position; all interossei muscles and flexors are involved, especially those of the fourth and fifth fingers. In hook grasp, all fingers are flexed, with the participation of the muscle flexor digitorum profundus. The *precision grasp* is divided into various types: palmar pinch (pad-to-pad), tip-to-tip pinch, lateral or key pinch, pencil apprehension, and interdigital or cigarette grasp [10, 11].

3.2 Locomotion and gait

Locomotion is the act of moving from place to place by means of one's own mechanisms or power [11]. The locomotion is realized: *on foot* (walking, running, ascending or descending stairs, jumping); *on wheels* (bicycling, skating, wheelchair); *on hands and knees* or *hands and feet*; *rotary locomotion* [11].

Gait is a manner of moving on foot [12]. Bipedal walking is an important characteristic of humans. The *gait cycle* is a repetitive pattern, including steps and strides. A step is one single footstep; a stride is the complete succession of one step with every lower extremity. The gait cycle is the sequence of one stance phase and one swing phase in proportion 3:2 (ratio 60 to 40%); the motion of every lower extremity is a combination of open-chain and closed-chain movements. Different sub-phases of the walking cycle are described [13–15]. Authors divide the stance phase into several sub-phases: heel strike (initial contact of the heel with the ground), foot flat (full contact of the foot with the base), mid-stance or support (support on one foot, with full weight bearing on this foot), heel-off, toe-off, and pre-swing (propulsion). The swing phase is divided into three sub-phases: acceleration (initial swing), mid-swing (constant velocity of the leg movement), and deceleration (late or terminal swing).

In clinical practice, we currently use several variables for gait analysis: step length, cadence (steps per minute), and walking velocity.

Pathological gait can be due to many musculoskeletal or neurological conditions. Every joint disorder of the lower limb may cause pathological gait: hip, knee, foot, and ankle pathology. Arthritis, muscle and joint contractures, leg length discrepancy, and pain can provoke the development of a pathological gait. Numerous neurological diseases can cause pathological gait types: hemiplegic gait, paraplegic gait, parkinsonian gait, and ataxic gait. The most frequent types of pathological gait in rehabilitation practice are presented in **Figure 1**. For example, spastic hemiparetic gait (Wernicke-Mann posture and gait – in patients with post-stroke hemiparesis), Parkinsonic gait, spastic paraparetic gait (in cases with multiple sclerosis, cerebral palsy, or after spinal cord injury at the thoracic level), gluteus medius gait (trendelenburg gait – in hip osteoarthritis), steppage gait (in peroneal paresis), antalgic gait (in discal hernia or in coxarthrosis), etc.

In rehabilitation, we use many qualitative methods for the evaluation of gait: Walking tests for 2, 6 minutes, or for 10 meters; timed up and go (TUG) test; and Tinetti test. For balance assessment, we apply the Romberg test (with open eyes or with closed eyes).

3.3 Grasp and gait rehabilitation

Grasp and gait (GG) are important activities, necessary for autonomy in everyday life. Therefore, grasp, balance, and gait recovery are significant indices for the efficacy of rehabilitation.

We use standardized protocols for recovery of the grasp, balance, and gait [11].

Before the initiation of the grasp training, we need some preparation by functional exercises: bilateral hand use; thumb opposition; in-hand manipulation; and pre-writing activities – make a puzzle and cut the line. Occupational therapists propose some sensory-motor games – with tennis balls, cotton balls, Lego, finger paint, stamps, stickers, and putty activities. Ultimately, several authors discuss the impact of *mirror box therapy* – for training different grasp types, especially in patients with hemiplegia or hemiparesis [11].

In post-stroke hemiparesis, we observe a characteristic position of the body and extremities, and spastic hemiparetic gait (Wernicke – Mann posture and gait). We apply position therapy; proprioceptive neuromuscular facilitation (PNF); electrical stimulations for spasticity control and improvement of muscle weakness; cryotherapy

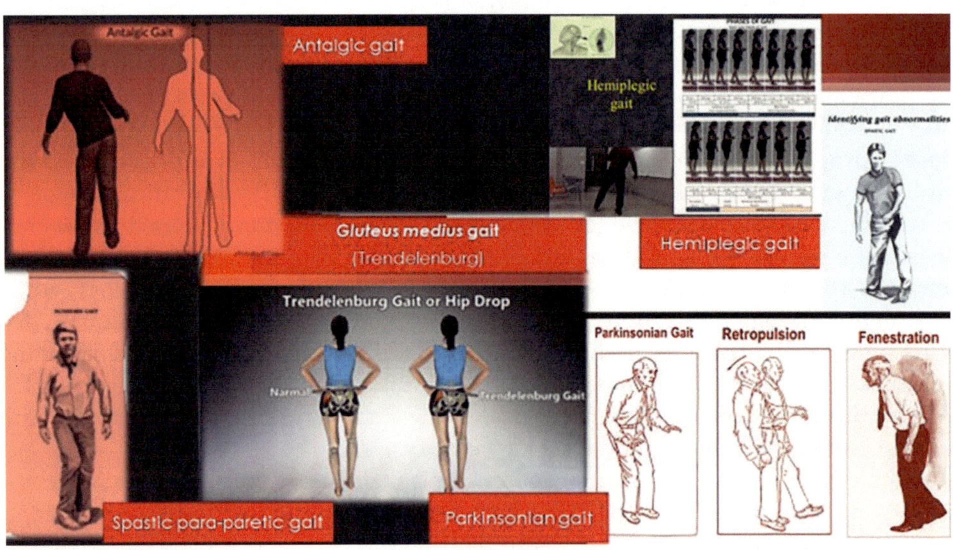

Figure 1.
Frequent pathological gait in rehabilitation.

or paraffin applications; grasp, balance, and gait training. In case of hemiparetic shoulder, we use interferential electric currents, low-intensity low-frequency magnetic field, functional electrical stimulations of the central fibers of the deltoid muscle, cryotherapy, and analytic exercises. In case of a developed hemiplegic hand, we apply magnetic field, laser therapy, paraffin or ice, active exercises, electrical stimulations of the antagonists of the spastic muscles, and mirror therapy.

The classical algorithm in multiple sclerosis includes magnetic field, cryotherapy, hydrotherapy with low temperature of the water (under the indifferent level), balneotherapy with mineral waters (containing sulphur), and underwater exercises.

In Parkinsonism, we use interferential currents, magnetic field, relaxing massage, physiotherapy, and occupational therapy.

In peripheral neurological conditions (as lumbo-sacral radiculopathy), we adapt the algorithm to the phase of the disease. During the acute stage, we apply diadynamic electric currents, magnetic field, transcutaneous electro-neuro-stimulation (TENS), deep oscillation (DO), relaxing massage, analytic exercises, and position therapy. In the sub-acute stage, we prefer TENS, middle-frequency electric currents (interferential currents, sinusoidal-modulated currents, Kotz currents), soft-tissue techniques, and active exercises. During the chronic stage, we include high-frequency electric currents, ultra-sound or phonophoresis with non-steroidal anti-inflammatory drugs, massage, exercises, peloids (therapeutic mud applications, sea lye compresses), balneotherapy, underwater massage, and underwater exercises. In cases with peripheral paresis, the algorithm recommends electrophoresis (iontophoresis) with galantamine hydrochloride (Bulgarian drug: Nivalin) and functional electrical stimulations (with exponential form of the pulses) in the motor points of the altered nerve and of the paretic muscles, analytic exercises, and underwater gymnastics. We consider most effective the gait training in water.

In orthopedic and traumatic disorders, the algorithm for gait recovery includes interferential currents, magnetic field, electrical stimulations, cryotherapy, and exercises. In bone fractures, we apply interferential currents, magnetic field, position therapy, and exercises for muscle force. In joint luxation, we prescribe interferential electric currents, magnetic field, TENS, diadynamic currents, cryotherapy, and exercises for range of motion.

In rheumatologic conditions, we use TENS, low and middle frequency electric currents, electrical stimulations, DO, soft tissue techniques, and active exercises. In degenerative joint conditions, we can use thermotherapy and peloids. In inflammatory joint diseases, during the acute stage we apply cryotherapy.

In every case, the basic rehabilitation algorithm must be adapted to the individual patient's needs.

4. Robots, virtual reality, and robotic rehabilitation

4.1 Robot

The origin of the term "Robot" is from the proto-slavic word "robota" or "rabota," meaning work or hard work in many Slavic languages, as Czech, Slovak, Bulgarian, Russian, Polish, Ukrainian, and others [16].

The Robot Institute of America defined a robot as "a programmable, multi-functional manipulator designed to move material, parts or specialized devices through variable programmed motions for the performance of a variety of tasks" [17].

The *three Ds "Dull, Dirty, and Dangerous"* describe the tasks of a robot (with repetitive movements and weight-support of patients in rehabilitation clinical practice) [18]. Robots help the staff in the process of control and measurements of patients' movements [19].

Different types of robots are applied in neurorehabilitation: for upper and for lower extremities, unilateral (for grasp training) and bilateral (for gait training); exoskeletons and controllers of endpoint trajectories.

We must underline the difference between robots and electro-mechanical devices, (for example, treadmill with body weight support); the presence of "intelligent" sensors is the key point for differentiation [20].

4.2 Neurorobotics

Neurorobotics is the branch of science using neuroscience, robotics, and artificial intelligence [21].

Neurorobots have the capacity to adapt, due to their on-board sensors. The treadmill with body-weight support is an electro-mechanical device and not a robot [22]. On the other hand, Lokomat (a device for gait training) is a neurorobot, using sensors for adaptation of its function to the patient's performance [17, 23].

The safety of the human beings is the most important in the construction of the robots. The three laws of robotics are firstly explained in the book of Isaac Asimov "I, Robot" in 1950 [24]. During the application of neurorobots, we need safety and efficacy. The formulation of the three laws of neurorobotics includes these requests [23]. The first law for safety of medical devices and robots and the need for a high *Benefit/Risk* ratio are supported by the requirements of the International Organization for Standardization [25, 26].

Many ethical dilemmas can be generated in the practice of robotic NR. We must clarify the ethical issues to constructors and therapists [27, 28].

The capacity of the artificial intelligence to support human intelligence is enormous [28, 29]. The main qualities of neurorobots for NR include high mechanical compliance, large range of force, adaptive assistance properties, possibility to assess the functional potential of the patient and to evaluate the progress of the functional level; flexibility (capacity for individualized treatment); and uniformity (sufficient scientific evidence), etc. [21, 30, 31].

Effectiveness of neurorobots in rehabilitation is proved by numerous studies on various topics, especially in patients after traumatic spinal cord injury, and on NR in post-stroke hemiparesis [32, 33].

Some limitations of NR robots are described: monotony, exacerbation of spasticity, economic barriers, etc. [21]

4.3 Virtual reality

In numerous sources, the term virtual reality (VR) is used for any computer-based technology providing monitor-based visualization. Nevertheless, the strict definition must include a user-computer interface, generating an environment including many sensorial channels (visual, auditory, and even smell and taste). A standard videogame is not a VR (for the moment). VR applies the *three Is: immersion, interaction,* and *imagination* [34].

Ultimately, many studies were published, proving efficacy of application of VR in neurorehabilitation, especially for the recovery of patients' autonomy in activities of daily living, emphasizing grasp, balance, and gait [35–40].

5. Robotic neurorehabilitation and neuroplasticity

5.1 Robotic NR

The goal of neurorehabilitation is to stimulate functional recovery by therapeutic interventions, oriented to excite the activity-dependent or experience-dependent neuroplasticity [10, 32].

Traditionally, NR uses methods and techniques of physical medicine: natural and preformed physical modalities, such as physiotherapy, occupational therapy, underwater exercises, soft-tissue techniques, electrical stimulations, TENS, magnetic field, laser, DO, etc. During the past few years, we observe the introduction of new methods in NR practice: constraint-induced movement therapy (intensive, experience-based, repetitive motor training of a paretic limb); weight-supported treadmill walking (intensive, experience-dependent functional movement training); motor training for language recovery; prism adaptation training for spatial neglect (with virtual feedback); transcranial magnetic stimulation; and transcranial direct current stimulation [19, 41, 42].

Principles of NR include task-specific and goal-oriented practice, increasing difficulty, multisensory stimulation, rhythmic cueing, knowledge of performance and of results, and social interaction [32, 43].

Many modern devices and techniques are applied ultimately in the clinical practice of NR, especially robotics (neurorobotics) and virtual reality [44]. The future of NR for post-stroke patients includes the application of modern devices, such as robots, brain-computer interfaces, noninvasive brain stimulators, neuroprosthesis, virtual reality, wearable devices, tablet PC, and others [20].

5.2 Neuroplasticity

According to the medical dictionaries: "neuroplasticity is the brain's capacity to reorganize itself by forming new neural connections throughout life." The "aim" of neuroplasticity is to optimize neural networks during phylogenesis, ontogenesis and physiological learning, and in case of brain disease [45, 46].

Principal mechanisms of brain repair are based on brain plasticity (spontaneous recovery, input of "axonal sprouting" and "mirror-neurons", use-dependent plasticity, synaptic or gray matter plasticity, and white matter plasticity) [47]. Many factors influence neuroplasticity, for example, physical activity, exercises, diet, and rehabilitation procedures [48–51].

The goal of NR is the patient's adaptation to the "new" situation (of neuronal alteration).

Neuroplasticity is the pathophysiological basis for treatment of the cerebral lesions through physical training and rehabilitation, including goal-directed activities [10]. In practice, every rehabilitation process uses neuroplasticity *(training plasticity, use-dependent* or *activity-depending plasticity)* for functional recovery and amelioration of independence in the everyday life of neurological patients [52, 53].

6. Our clinical experience

We have clinical experience with patients suffering from different neurological and neurosurgical conditions: hemiparesis after cerebral vascular accident, multiple sclerosis, Guillain-Barre syndrome and after neurosurgical intervention for brain or spinal cord aneurysm, tumor, or trauma. Here, we present some of our own results with post-stroke hemiparesis and after spinal cord injury.

6.1 Robotic NR in post-stroke hemiparesis

The goals of stroke rehabilitation are to maximize functional independence; to prevent and manage comorbidities; to optimize psycho-social adaptation of patient and family; to enhance quality of life; to facilitate resumption of prior life roles; and to reintegrate patient in professional life and in social life.

The functional recovery after a cerebro-vascular accident depends on the three Ps: Preservation, Prevention and Plasticity (early treatment for preservation of the cerebral tissue, prevention of the brain loss and of eventual recurrent stroke, using plasticity through rehabilitation). The brain possesses a significant capacity to reorganize itself and to recover from loss of function, following a stroke. Rehabilitation training stimulates brain reorganization and accelerates functional recovery.

NR in post-stroke patients traditionally applies different PRM techniques: proprioceptive neuromuscular facilitation, Kabath and Brunnstrom methods, neurodevelopmental training (Bobath concept), and sensorimotor therapy (Rood approach). Ultimately, some modern modalities were included in the rehabilitation process, such as motor relearning program, constraint-induced movement therapy, mirror therapy, transcranial electric (direct current) stimulation, transcranial magnetic stimulation, low intensity focused transcranial ultrasonic stimulation, functional electrical stimulation (FES), vagus nerve stimulation, electromyographic (EMG) biofeedback, robotic devices, and virtual reality.

6.1.1 Clinical case of NR with exoskeleton in post-stroke hemiparesis

Ultimately, we observe an increasing frequency of cerebro-vascular diseases (with high levels of mortality and disability) in several countries, including Bulgaria.

We present a clinical case of a female patient, 67 years old, 3 months after an ischemic stroke.

For evaluation, we use detailed neurological exam, functional assessment through classical scales and the International Classification of Functioning (ICF). At the entry, the patient presented left spastic hemiparesis, motor functions level 3–4 according the Brunnstrom classification, left superficial hemi-hypoesthesia, and contractures of the left wrist and fingers, ankle, and toes.

For treatment, we applied a complex NR program (of 20 procedures), including traditional physiotherapy and ergotherapy procedures, balance and gait training, electrical stimulations for the muscles – extensors of the wrist, ankle, and toes of the paretic extremities (with tetanic pulses), and robotic NR with exoskeleton (a Cyberdyne device) for balance and gait training (**Figure 2**). According to dictionaries, an *exoskeleton (from Greek ἔξω, éxō "outer" and σκελετός, skeletos "skeleton")* is the external skeleton that supports and protects the body, in distinction with

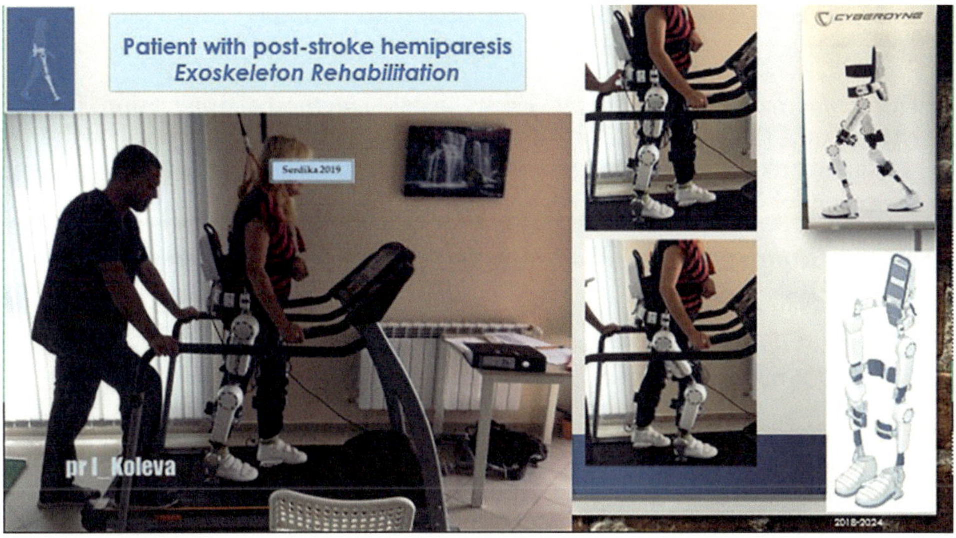

Figure 2.
Gait training of a patient with post-stroke hemiparesis with exoskeleton.

the internal skeleton (endo-skeleton). In this case, we used the hybrid assistive limb (HAL) – exoskeleton for lower extremities (double leg), medium size, and leg length – size M.

We observed significant functional recovery: reduction of the muscle weakness and contractures, stabilization of the equilibrium, retrieval of the autonomic gait (with one cane), and amelioration of autonomy in activities of daily living. Therefore, the NR with exoskeleton improves the bipedal synchronization and ameliorates the "normal" gait pattern of the patient [54].

6.1.2 Robotic-based GG-NR of hemiparetics after stroke

6.1.2.1 NR program

During the NR process, we accentuate on functional recovery, capacities for different activities, and autonomy in everyday life. We apply a synergic combination of physiotherapy, occupational therapy, thermo- or cryotherapy, electrotherapy, etc. The duration of the NR course in our country traditionally is 20 procedures, 5 days a week, for 4 weeks. The PRM program includes:

- *Physiotherapy*: *individualized movement program* (proprioceptive neuro-muscular facilitation/PNF/techniques by Kabath or Bobath, analytic exercises for the paretic limbs, balance training, training of transfers and gait, exercises to stabilize the gait on flat terrain and on stairs; soft tissue techniques); duration 45 minutes, twice a day;

- *Cryotherapy* with ice block – distally in upper and lower extremities, duration 2–3 minutes per field, twice daily;

- *Occupational therapy (OT)*, including training in activities of daily living (ADL), especially for self-care – 30 minutes, twice a day;

- *Analytic exercises and manipulative activities (grasp training) with a mirror box* (mirror therapy) – 10–20 minutes a day;
- *Functional electrical stimulations* (FES) – a stable method of application, for the extensors of the paretic upper and lower limbs (with special attention to the wrist and ankle) and for the flexors of the fingers and toes, with tetanic pulses, a procedure of 10–20 minutes, once a day;
- *Training of the upper limb with tyro-motion devices* – 30–60 minutes a day (see **Figure 3**);
- *Training of the paretic lower limb with Lokomat Pro (Hocoma device)* – a procedure of 45–60 minutes, every day (see **Figure 4**);
- *Patient education* (position therapy, control of spasticity, grasp and gait training; hypolipid and hypoglucid diet; ending of smoking and alcohol abuse; control of lipid profile and arterial hypertension; and stress control).

6.1.2.2 Results of the robotic-based grasp-NR

Figure 3 presents the potential of the Tyro-motion station for training different types of grasp – cylindrical, precise, interdigital, etc.

The tyro-system has the capacity to realize quantitative evaluation of the grasp recovery during consecutive sessions (**Figures 5–8**). In different patients, we observed amelioration of the grasp and the autonomy in activities of daily living – ADL (**Figure 5**), of the joint range of motion (**Figure 6**) and of muscle force (**Figure 7**). The reduction of spasticity in the paretic upper extremity was evaluated using two spasticity scales – of Tardieu and of Ashworth (**Figure 8**).

During past few years, we had the possibility to realize a comparison between traditional NR and robotic NR [54]. The NR program in all patients with post-stroke

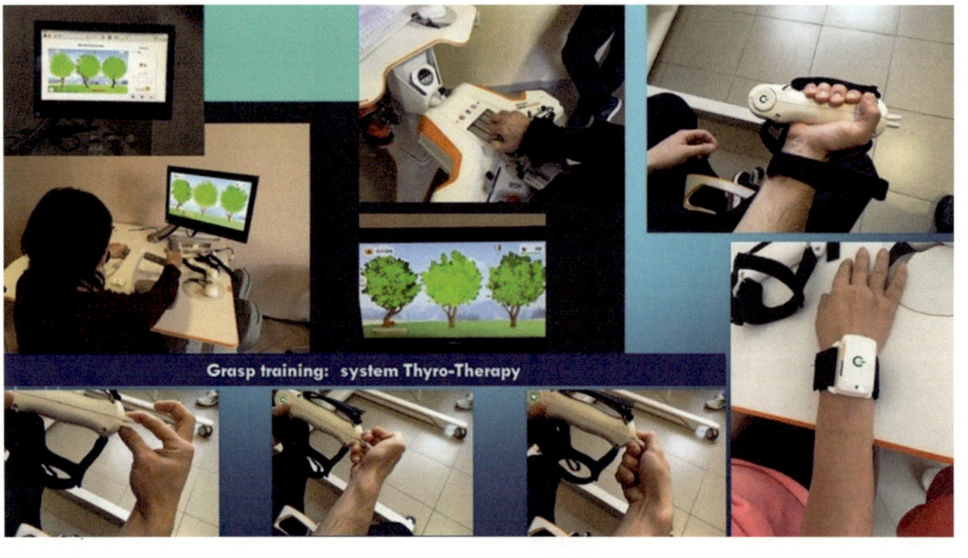

Figure 3.
Grasp and grip training of post-stroke patients with tyro-station.

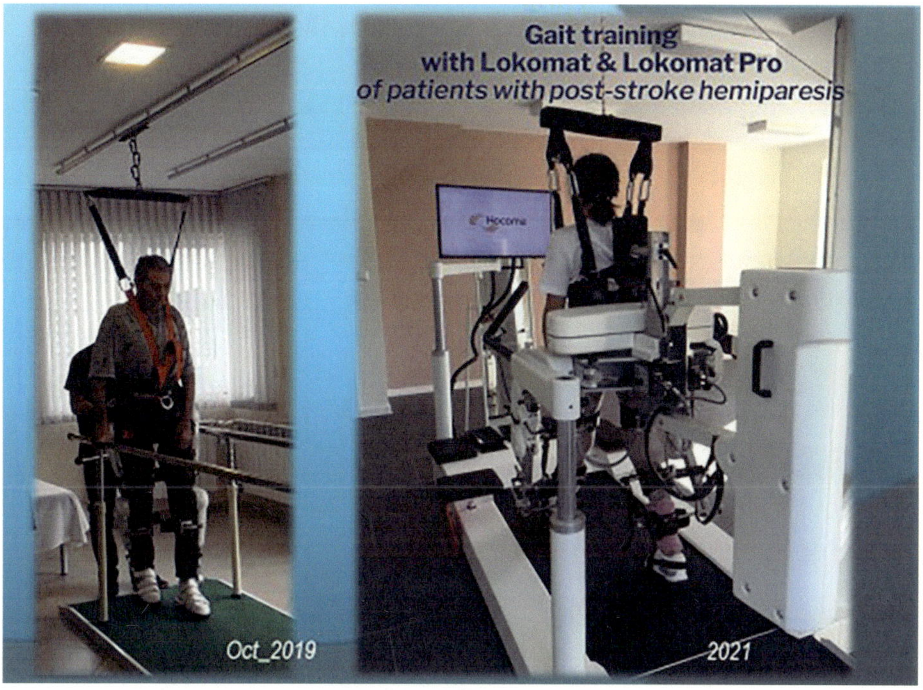

Figure 4.
Gait training of patients with post-stroke hemiparesis.

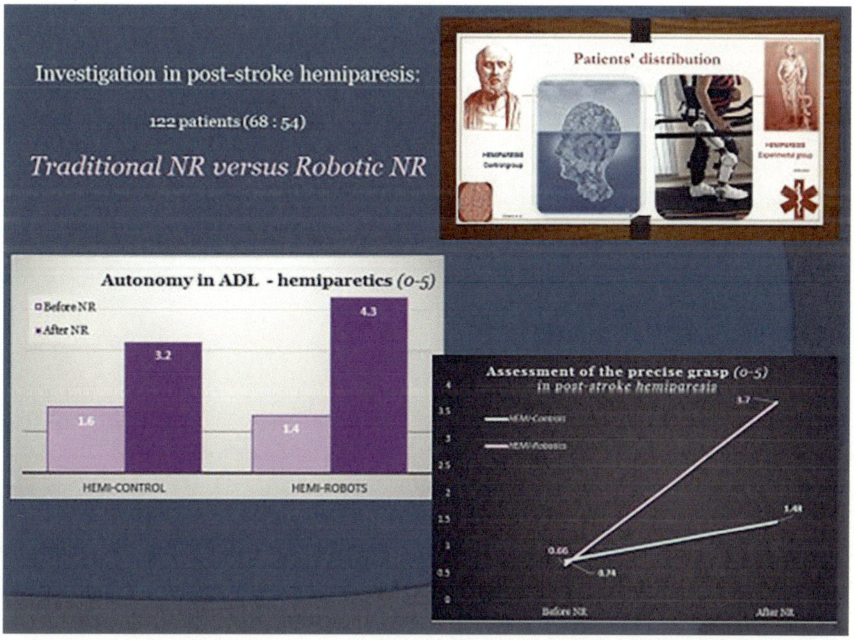

Figure 5.
Traditional NR versus robotic NR – Our own results for upper limb.

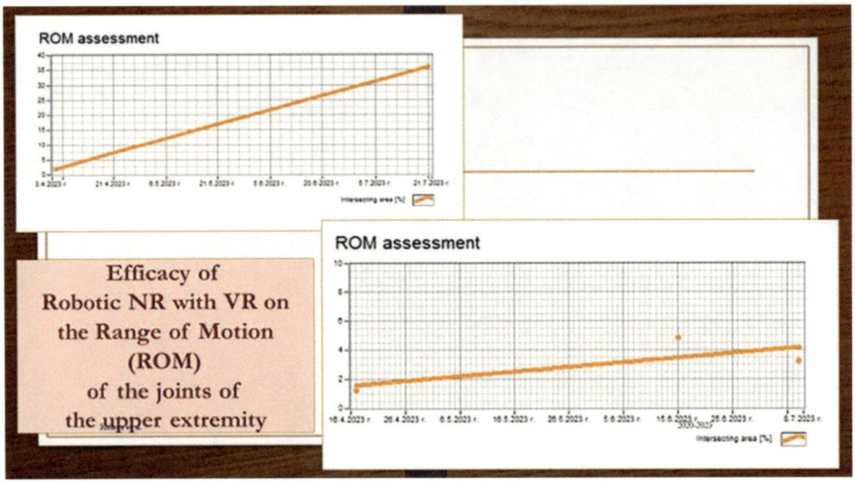

Figure 6.
Efficacy of robotic NR on the range-of-motion of the upper extremity.

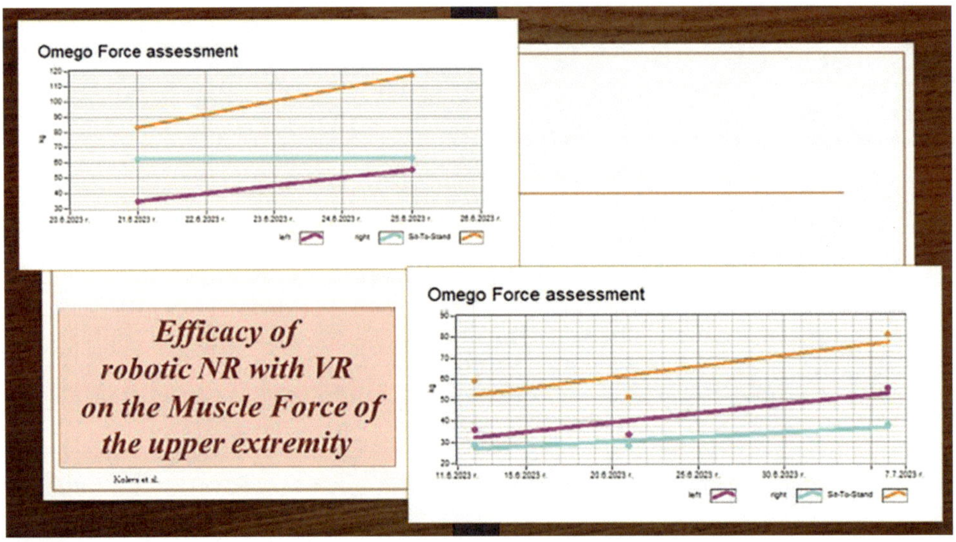

Figure 7.
Efficacy of robotic NR on the muscle force of the upper extremity.

hemiparesis included: traditional physiotherapy, occupational therapy, cryotherapy, and functional electrical stimulations. The control group (68 patients) received only these procedures. In the experimental group (54 patients), we added robotic NR – three sessions weekly. We obtained a significant increase of the force of the precise grasp with the paretic hand and amelioration in autonomy in activities of daily living (ADL), as presented in **Figure 5**.

6.1.2.3 Results of the robotic-based gait-NR

The Lokomat system of Hocoma is designated for intensive functional gait training, including a computer-guided electric orthosis for assistance of movements

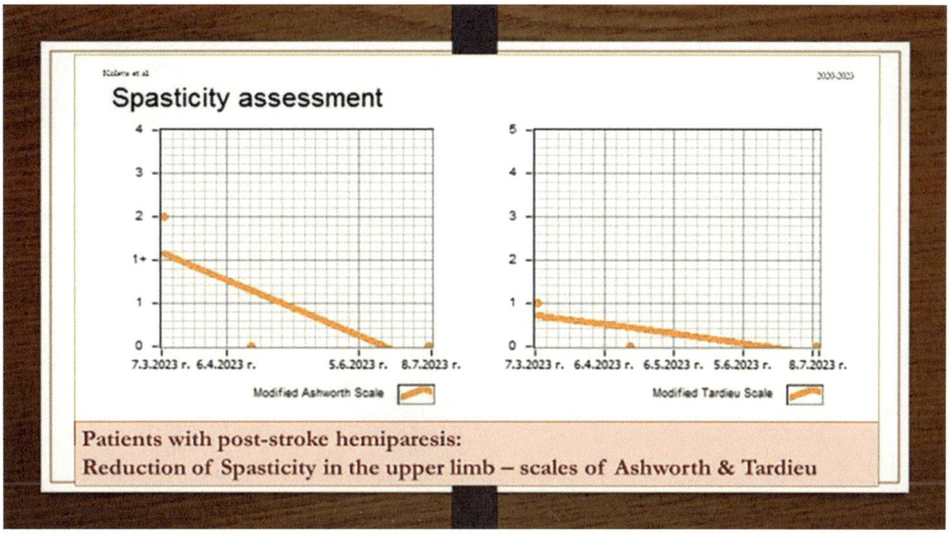

Figure 8.
Reduction of spasticity in the upper limb muscles during robotic NR – Using both scales (of Ashworth and Tardieu).

of lower extremities during locomotion (**Figure 4**). The device Lokomat Pro (professional) has the potential of regulation of the speed, load and robotic aid, and can be adapted to the patient's measures and physical capacity. The exoskeleton of Lokomat Pro corrects the gait pattern. The inclusion of elements of virtual reality (audio-visual stimuli, as environment, music, games) ameliorates the patient's tolerance, especially in children or in older adults.

Principal indications include motor weakness due to neurological diseases and conditions (hemiparesis after stroke or traumatic brain injury, multiple sclerosis, Parkinsonism, cerebral palsy, paraparesis after spinal cord injury, Guillain-Barre syndrome, and peripheral paresis) and degenerative joint diseases (as hip and knee osteoarthritis). The device can be applied in patients with a height of maximum 200 cm and weight up to 135 kg.

Absolute contra-indications include osteopenia and osteoporosis, fractures of lower limbs, cardiac or respiratory insufficiency, cognitive deficiency, infections, epilepsy, and pregnancy. Relative contra-indications are arthroplasty, muscle or joint contractures and joint instability in lower limbs, loss of control of the head and the upper part of the body, skin lesions in the region of contact with the orthosis, and electronic devices (as intrathecal baclofen pump and pace-maker).

The device has the potential to reduce partially the weight in lower limbs and to correct the gait pattern. These capacities are very useful in cases of muscle weakness and pathologic gait (Wernicke-Mann gait in post-stroke hemiparesis, spastic paraparetic gait in spinal cord injury at thoracic level, etc.).

The Lokomat Pro system of Hocoma has the capacity to realize quantitative evaluation of the gait recovery during consecutive sessions.

Figures 9–12 present the results of the Lokomat-NR of different patients with hemiparesis after cerebrovascular accident. We compared parameters between the first and the last therapeutic session – usually our patients have 20–30 procedures of 50 minutes, for a period of 1 to 3 months, and frequency of sessions is twice a week. The number of procedures is adapted to the patients' potential and needs (physical condition and comorbidities). We observed an increase of the distance during the

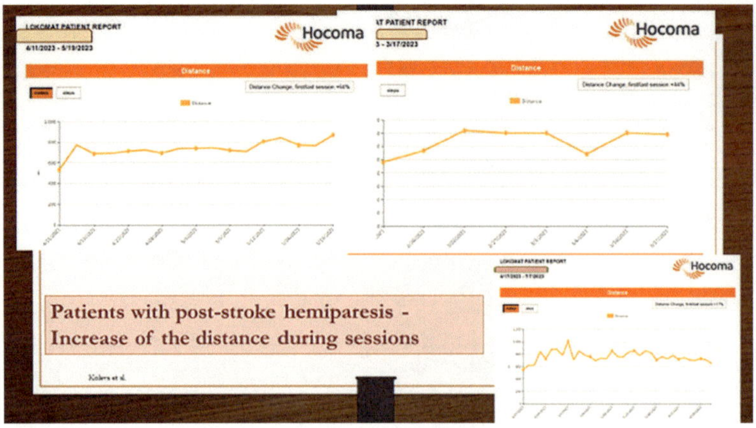

Figure 9.
Lokomat Pro results: Increase of the distance during sessions.

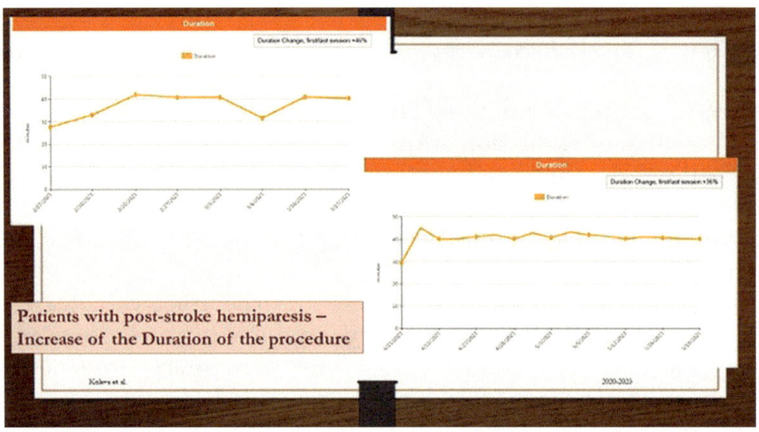

Figure 10.
Increase of the duration of the procedure with Lokomat Pro.

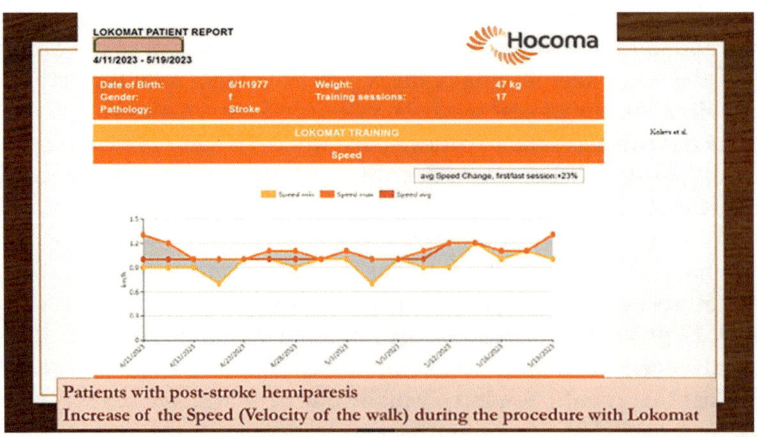

Figure 11.
Locomat Pro results: Increase of the gait velocity.

DOI: http://dx.doi.org/10.5772/intechopen.1004544

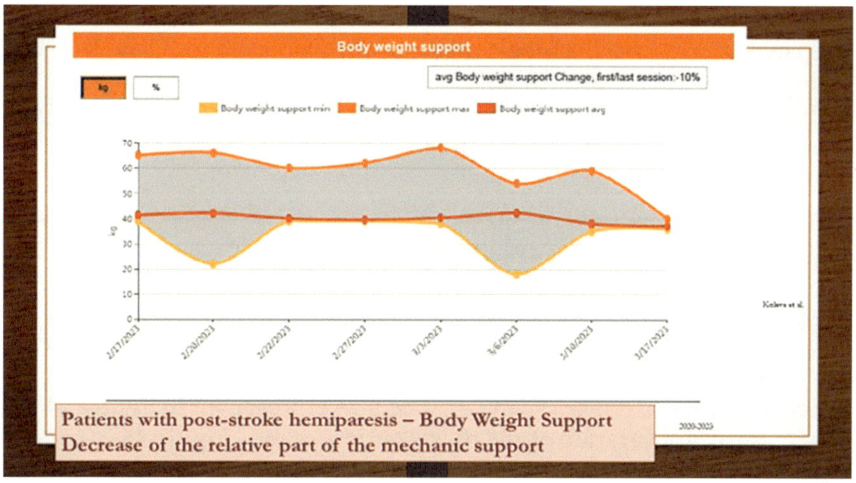

Figure 12.
Body weight support – Reduction of the relative part of the mechanic support.

therapeutic session (**Figure 9**), increase of the duration of the procedure (**Figure 10**), increase of the gait velocity (**Figure 11**), and changes in body weight support – reduction of the relative part of the mechanic support (**Figure 12**).

6.2 Lokomat NR after spinal cord injury with inferior paraparesis

Patients after spinal cord injury (SCI) present loss of motor control and loss of sensibility. In cases after SCI at thoracic level, we observe inferior spastic paraplegia. Entry at the NR department is usually 1 to 3 months after the neurosurgical stabilization (**Figure 13**).

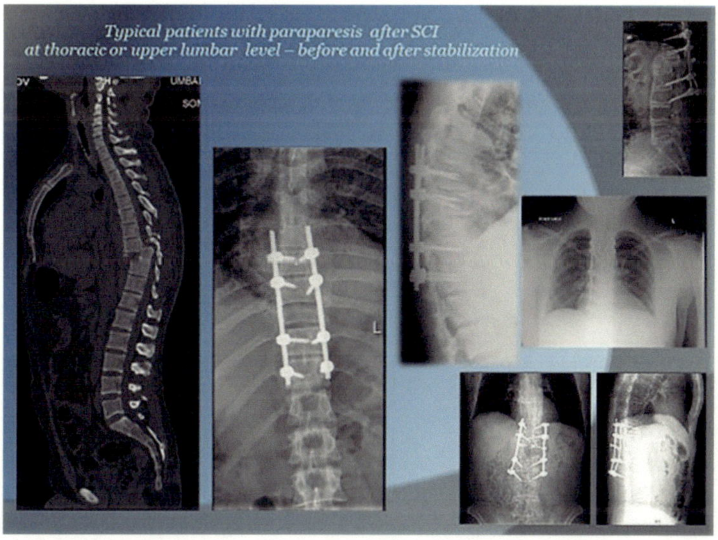

Figure 13.
X-rays of typical patients with inferior paraplegia/paraparesis.

We apply a standard NR program, including physiotherapy, ergotherapy, cryotherapy, electrical stimulations for extensors of the ankle and toes, and balance and gait training with Lokomat Pro (Hocoma system).

6.2.1 Results of the Lokomat-NR of paraparetics after SCI

The next figures present the results of the Lokomat-NR of a series of patients with inferior paraparesis after spinal cord injury at the thoracic level.

We took out the data directly from the device, and we received significant results, comparing different sessions of robotic NR: amelioration of functional parameters of the gait, increase of the range of motion and the muscle force of lower extremities, and improvement of the proprioception in lower limbs.

The increase of the distance during a procedure is presented in **Figure 14**.

The increase of the duration of the therapeutic session – **Figure 15**.

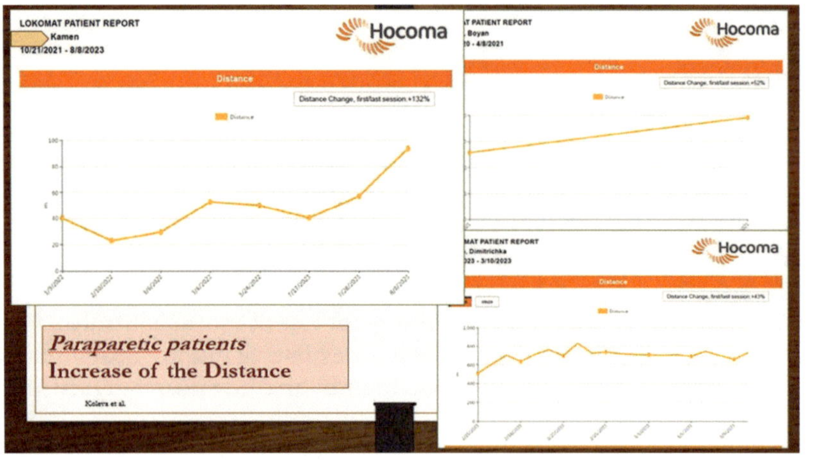

Figure 14.
Increase of the distance in paraparetics – Lokomat-NR.

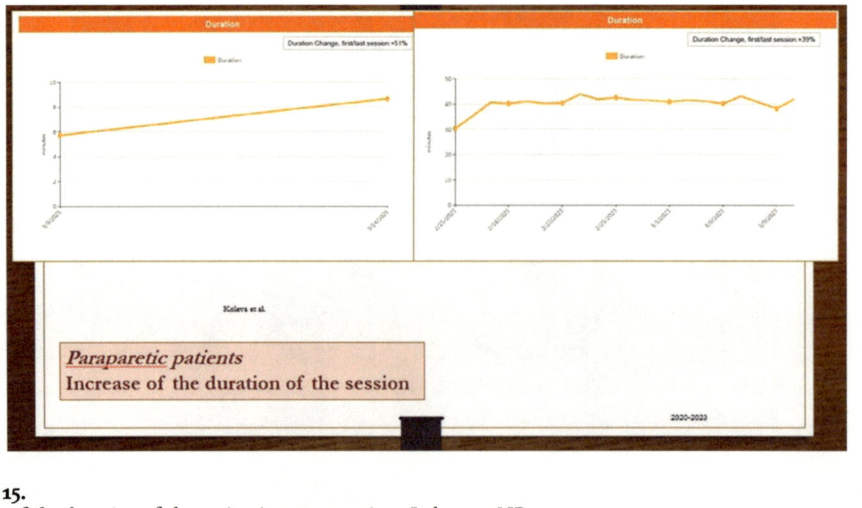

Figure 15.
Increase of the duration of the session in paraparetics – Lokomat-NR.

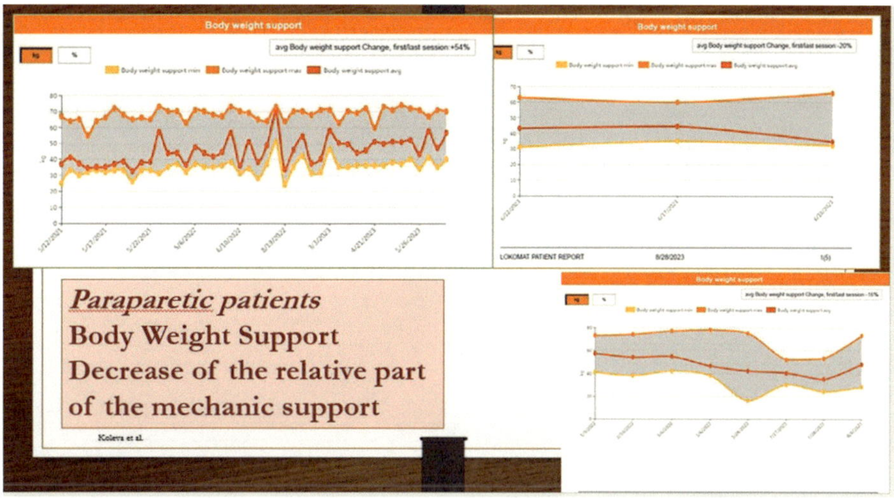

Figure 16.
Body weight support during sessions of Lokomat-NR.

The reduction of the mechanic support during sessions with a graphic of the dynamics in body weight support is presented in **Figure 16**.

Figure 17 demonstrates the increase of the range of motion in the lower limbs of paraparetics.

An amelioration of the proprioception is observed (**Figure 18**).

6.2.2 Traditional NR versus robotic NR – Our own results

During last years, we observed patients, treated with traditional NR and Lokomat-based NR. We noticed the difference in results, obtained in both types of rehabilitation methods – traditional NR and robotic NR [54].

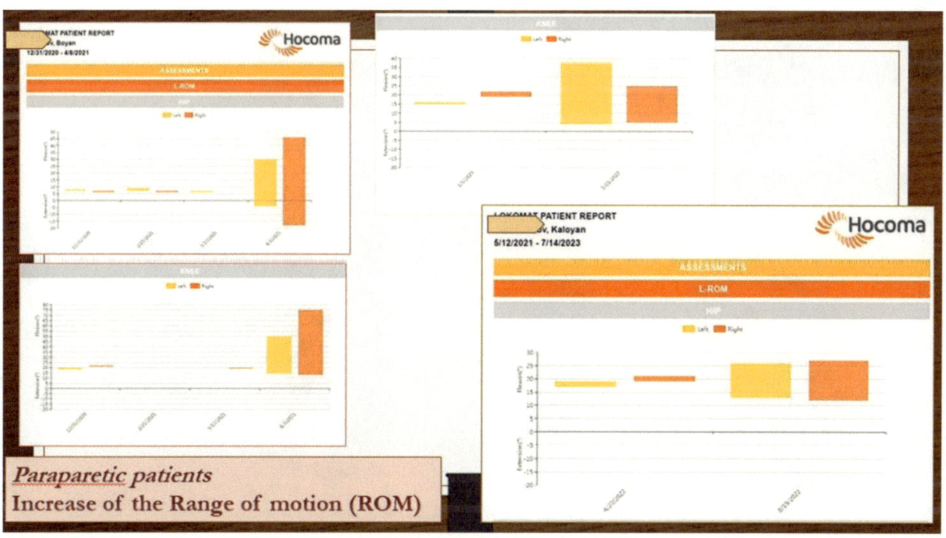

Figure 17.
Increase of the ROM in paraparetics – Lokomat-NR.

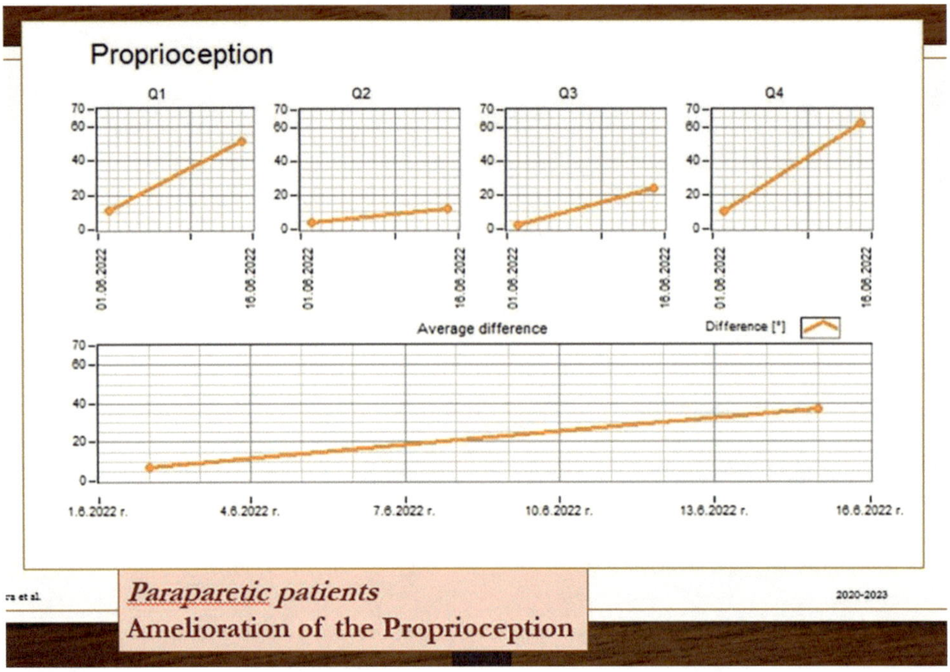

Figure 18.
Amelioration of the proprioception.

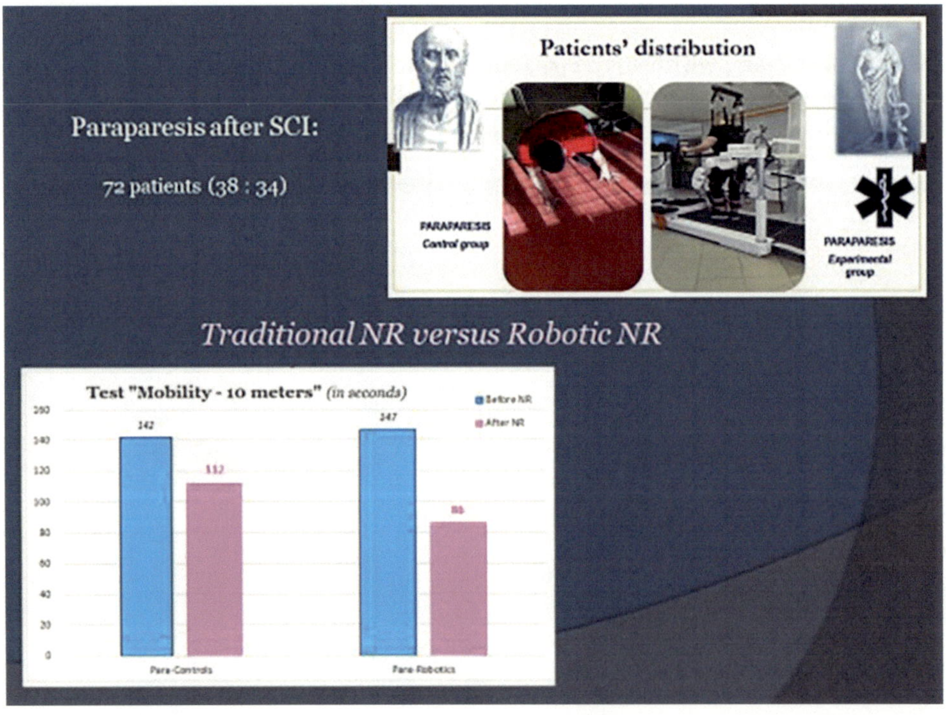

Figure 19.
Traditional versus robotic NR – Our own results for ambulation.

Here, we present some results, concerning the autonomic gait.

The NR program in all 72 patients with post-traumatic inferior spastic paraparesis included traditional physiotherapy, occupational therapy, cryotherapy, and functional electrical stimulations. The control group (38 patients) received only these procedures. In patients of the experimental group (34 patients), we added robotic NR – three sessions weekly.

We obtained an increase of the velocity of the gait (with technical aids – wheelchair, crutches, and canes) in both groups, but the results were most significant in the experimental group (with Lokomat training), as demonstrated in **Figure 19**.

7. Conclusion

The application of robotic rehabilitation accelerates the process of grasp and gait recovery, improves the quality of care, and ameliorates the autonomy in everyday life of neurologic patients (especially with post-stroke hemiparesis and inferior paraparesis after spinal cord injury).

Complex neurorehabilitation with robots and virtual reality stimulates activity-induced neuroplasticity, enhances bimanual and bipedal synchronization, and corrects the patterns of the grasp, balance, and gait.

Acknowledgements

There is no financial support for this research and work.

No ethical issues. We respect the requirements of the Declaration of Helsinki (1964) for the protection of patients. All patients signed written informed consent for every diagnostic and therapeutic procedure and for publication of the correspondent data and photos.

Conflict of interest

Authors declare no conflict of interest.

Author details

Ivet B. Koleva[1,2,3*], Radoslav R. Yoshinov[4] and Borislav R. Yoshinov[5]

1 Medical University of Sofia, Bulgaria

2 National Heart Hospital, Sofia, Bulgaria

3 Hospital for Long-term Care and Rehabilitation "Serdika" with Medical Center for Robotic Neurorehabilitation "ReGo", Sofia, Bulgaria

4 University for Library Studies and Information Technologies UNIBIT, Sofia, Bulgaria

5 Neurology Department, University Hospital for Active Care "St. Ivan Rislki", Sofia, Bulgaria

*Address all correspondence to: dr.yvette.5@gmail.com

References

[1] World Health Organization and the World Bank. World Report on Disability. Geneva: WHO Press; 2011

[2] World Health Organization. Classification of diseases. In: The ICD-10 Classification of Mental and Behavioral Disorders: Clinical Descriptors and Diagnostic Guidelines. 10th ed. Geneva: WHO Press; 1992

[3] World Health Organization. International Classification of Functioning, Disability and Health. Geneva: WHO Press; 2001

[4] Koleva I. Repetitorium physiotherapeuticum (basic principles of the modern physical and rehabilitation medicine). In: Book for English-Speaking Students. Sofia: SIMEL; 2008

[5] European PRM Bodies Alliance. White book on physical and rehabilitation medicine in Europe. European Journal of Physical and Rehabilitation Medicine. 2018;**54**(2):1-204

[6] Gutenbrunner C, Ward AB, Chamberlain MA. White book on physical and rehabilitation medicine. Journal of Rehabilitation Medicine. 2007;**1**(Supplement 45):1-48. Available from: www.medicaljournals.se/jrm [Accessed: Dec. 11, 2010]

[7] DeLisa. Physical Medicine and Rehabilitation – Principles and Practice. 4th ed. Philadelphia: Lippincott, Williams & Wilkins; 2005

[8] Melvin JL. Physical and rehabilitation medicine: Comments related to the white book on physical and rehabilitation medicine in Europe. European Journal of Physical and Rehabilitation Medicine. 2008;**44**(2):117-119

[9] Koleva I. Complex neurorehabilitation algorithms for functional recovery and amelioration of independence in everyday life in socially important invalidating neurological diseases [thesis] for Doctor-es-Sciences in physical and rehabilitation medicine; Sofia: Bulgaria. 2009

[10] Koleva I, Yoshinov R, Yoshinov B. Clinical Neurorehabilitation (Electronic Monograph). In: Koleva I, editor. Sofia: SIMEL; 2019. pp. 1-678. ISBN: 978-619-183-085-5

[11] Koleva I, Avramescu ET, Yoshinov R et al. Grasp and Gait Rehabilitation. In: Koleva I, Avramescu ET, editors. 1st ed. Sofia: SIMEL PRESS; 2017. pp. 1-394

[12] merriam-webster.com. 2010 [Online]. Available from: http://www.merriam-webster.com/dictionary/gait [Accessed: Dec. 10, 2015]

[13] Simoneau GG. Kinesiology of walking. In: Kinesiology of the Musculoskeletal System: Foundation for Physical Rehabilitation. St. Louis, Missouri: Mosby; 2002. pp. 523-569

[14] Hamilton N, Luttgens K. Kinesiology, Scientific Basis of Human Motion. Chapter 19. 10th ed. Madison, WI: Brown & Benchmark; 2002. pp. 467-494

[15] Floyd RT. Manual of Structural Kinesiology. 17th ed. Boston - New York - San Francisco - St. Louis - Bogota - Caracas - Lisbon - London - Madrid - Mexico City - Milan - Montreal - New Delhi - Santiago - Seoul - Singapore - Sydney - Taipei - Toronto: McGraw Hill Higher Education; 2009. pp. 1-416

[16] Oxford English Dictionary/etymology. Oxford English Dictionary. Available from: [Accessed: May 05, 2020]

[17] Xie M. Fundamental of Robotics: Linking Perception to Action. New Jersey - London - Singapore - Hong Hong Kong: World Scientific; 2003. pp. 1-718. ISBN: 981-238-313-1 - ISBN: 981-238-335-2

[18] Lin P, Abeny K, Bekey GA. Robot Ethics: The Ethical and Social Implications of Robotics. Cambridge: The MIT Press; 2012

[19] Huang V, Krakauer J. Robotic neurorehabilitation: A computational motor learning perspective. Journal of Neuroengineering and Rehabilitation. 2009;**6**(1):1-13

[20] Iosa M, Morone G, Fusco A, et al. Seven capital devices for the future of stroke rehabilitation. In: Stroke Research and Treatment. Hindawi Publishing Corporation; 2012. pp. 1-9

[21] Benedek I, Vanta O. Neurorobotics for health practitioners. Available from: www.efnr.org. [Accessed: Jan 13, 2023]

[22] Mehrholz J, Thomas S, Elsner B. Treadmill training and body weight support for walking after stroke. Cochrane Database of Systematic Reviews. 2017:1-185

[23] Iosa M, Morone G, Cherubini A, Paolucci S. The three Laws of Neurorobotics: A review of what Neurorehabilitation robots should do for patients and clinicians. Journal of Medical and Biological Engineering. 2016;**36**:1-11

[24] Asimov I. Robot. New York: Gnome Press; 1951

[25] International Organization for Standardization. Medical Devices: Application of Risk Management to Medical Devices. ISO, 14971. 2020.

[26] International Organization for Standardization. Robots and Robotic Devices: Safety Requirements for Personal Care Robots. ISO; 2014

[27] Morasso P, Casadio M, Gianoni P, et al. Desirable features of a "humanoid" robot-therapist. In: Proceedings of Annual International Conference of the IEEE Engineering in Medicine and Biology Society. 2009. pp. 2418-2421

[28] Datteri E. Predicting the long-term effects of human-robot interaction: A reflection on responsibility in medical robotics. Science and Engineering Ethics. 2013;**19**(1):139-160

[29] Hocoma Users' Manual. Available from: https://knowledge.hocoma.com [Accessed: Apr. 01, 2023]

[30] Krichmar JL, Hwu TJ. Design principles for neurorobotics. Frontiers in Neurorobotics. 2022. DOI: 10.3389/fnbot.2022.882518

[31] Janjarasjitt S. Methods and applications in neurorobotics. Frontiers in Neurorobotics. 2022. DOI: 10.3389/fnbot.2022.1111877

[32] Maier M, Ballester BR, Verschure PFMJ. Principles of Neurorehabilitation after stroke based on motor learning and brain plasticity mechanisms. Frontiers in Systems Neuroscience. 2019;**13**:1-18

[33] Hidler J, Nichols D, Pelliccio M, et al. Multicenter randomized clinical trial evaluating the effectiveness of the Lokomat in subacute stroke. Neurorehabilitation and Neural Repair. 2009;**23**(1):5-13

[34] Burdea GC, Coiffet P. Virtual Reality Technology. 2nd ed. Hoboken: John Wiley & Sons; 2003

[35] Holden MK. Virtual environments for motor rehabilitation: Review. Cyber Psychology and Behavior. 2005;**8**(3):187-211

[36] Holden MD, Dyar T. Virtual environment training: A new tool for neurorehabilitation. Neurology Report. 2002;**26**:62-71

[37] Piron L, Paolo T, Piccione F, et al. Virtual environment training therapy for arm motor rehabilitation. Presence. 2005;**14**:732-740

[38] Lange B, Chang CY, Suma E, et al. Development and evaluation of low cost gamebased balance rehabilitation tool using the Microsoft Kinect sensor. In: Proceedings of the Annual International Conference of the IEEE Engineering in Medicine and Biology Society. 2011. pp. 1831-1834

[39] Laver KE, George S, Thomas S, et al. Virtual reality for stroke rehabilitation. Cochrane Database of Systematic Reviews. 2011;7:9. Article ID: CD008349

[40] Laver KE, Lange B, George S, et al. Virtual reality for stroke rehabilitation. Cochrane Database of Systematic Reviews. 2017;**11**(11):1-164

[41] Barrett AM, Oh-Park M, Chen P, Ifejika NL, editors. Neurorehabilitation - five new things. In: Neurology Clinical Practice. American Academy of Neurology; Dec 2013. pp. 484-491

[42] Yang YW, Pan WX, Xie Q. Combined effect of repetitive transcranial magnetic stimulation and physical exercise on cortical plasticity. Neural Regeneration Research. 2020;**15**(11):1986. DOI: 10.4103/1673-5374.282239

[43] Barbeau H. Locomotor training in neurorehabilitation. Emerging rehabilitation concepts. Neurorehabilitation and Neural Repair. 2003;**17**(1):1-11

[44] Holden MK, Dyar TH. Virtual environment training: A new tool for Neurorehabilitation. Department of Brain and Cognitive Sciences. 2002;**26**(2):62-71

[45] Ilg W, Synofzik M, Brotz D, et al. Intensive coordinative training improves motor performance in degenerative cerebellar disease. Neurology. 2009;**73**(22):1823

[46] Shaffer J. Neuroplasticity and clinical practice: Building brain power for health. Frontiers in Psychology. 2016;7:1118. DOI: 10.3389/fpsyg.2016.01118

[47] Kramer AF, Erickson KI. Capitalizing on cortical plasticity: Influence of physical activity on cognition and brain function. Trends in Cognitive Sciences. 2007;**11**(8):342-348

[48] Phillips C. Lifestyle modulators of neuroplasticity: How physical activity, mental engagement, and diet promote cognitive health during aging. Neural Plasticity. 2017. DOI: 10.1155/2017/3589271

[49] Vaynman S, Gomez-Pinilla F. Revenge of the "sit": How lifestyle impacts neuronal and cognitive health through molecular systems that interface energy metabolism with neuronal plasticity. Journal of Neuroscience Research. 2006;**84**(4):699-715

[50] De Oliveira Bristot VJ, De Bem Alves AC, Cardoso LR, et al. The role of PGC-1α/UCP2 Signaling in the beneficial effects of physical exercise on the brain. Frontiers in Neuroscience. 2019:13. DOI: 10.3389/fnins.2019.00292

[51] Moore D, Loprinzi PD. Exercise influences episodic memory via changes in hippocampal neurocircuitry and

long-term potentiation. European Journal of Neuroscience. 2020. DOI: 10.1111/ejn.14728

[52] Didier J. La plasticité de la fonction motrice. Collection de l'Académie Européenne de Médecine de Réadaptation. Paris: Springer Verlag; 2004. pp. 1-476

[53] Koleva I, Yoshinov BR, Yoshinov RR. Balance and gait neurorehabilitation in multiple sclerosis with significant fatigue and spasticity. Neurophysiology Research. 2020;**2**:1-7

[54] Zvetkova N, Yoshinov B, Koleva I. Robotic Neurorehabilitation. In: Koleva I, editor. Sofia: SIMEL; 2023 [In Bulgarian]

Chapter 8

Value Added Care: Improving Quality with Decreased Costs in Robotic Assisted Colorectal Surgery

Alexandra Green, Ashley Alexander and Bethany Malone

Abstract

Diseases of the colon and rectum, such as diverticulitis or cancer, are increasing and occurring in younger populations. Although robotic assisted surgery can be more expensive for the index operation, robotic assisted colorectal surgery decreases length of stay, need for narcotics, and the amount of time needed off work for recovery. Robotic surgery also introduces new opportunities for simulation training which allows developing surgeons to improve their technique without risk to a live patient. These benefits contribute to decreased healthcare costs for the system as a whole as well as mitigate downstream effects on the economy due to loss of productivity and tax revenue. Robotic assisted surgery also facilitates video-based mentorship. Video mentorship can be used to improve surgical care and health equity nationally and internationally. Robotic surgery may also provide future opportunities for integration of CT scans into the operative field or artificial intelligence to improve surgeon performance and accuracy.

Keywords: robotic surgery, robotics, colon and rectal surgery, colon cancer, rectal cancer, minimally invasive surgery, healthcare costs, healthcare economics

1. Introduction

The term "Robot" is defined in Merriam-Webster Dictionary as "a machine that resembles a living creature in being capable of moving independently and performing complex actions". According to the modern definition, "robotic surgery" would seem to suggest that the technology functions autonomously, which is not accurate. "Robotic surgery" is the term typically referring to the Da Vinci surgical platform used today in various surgical procedures. It would more precisely be described using the term "telepresence", meaning "transporting one's awareness to a different location" [1].

Robotic surgery first originated in 1985 with a device called PUMA used to position a stereotactic brain biopsy needle while reducing surgeon tremor and increasing precision. This device evolved into other similar devices used for procedures such as transurethral resection of the prostate, hip replacements, and percutaneous kidney biopsies. The current iteration of robotic surgery began as a Department of Defense and NASA initiative to allow surgeons to begin a life-saving operation while the patient was still being transported from a hostile or unsafe environment with the intention of military use on the battlefield. The first prototype in 1986, called "telepresence surgical system", which initially targeted microsurgery, evolved over time to address laparoscopy and laparoscopic-specific issues including reducing fulcrum effect, tremor, improving dexterity, and stereoscopic vision. After training simulation and animal model testing, Intuitive Surgical acquired the intellectual property in 1995. Intuitive developed their next generation, da Vinci, and began human trials in 1998. Meanwhile, the first FDA-approved robotic surgical system developed by Computer Motion comprised of a voice-controlled laparoscope. In 2003, after several years of further developments including robotic arms, table positioning, and interchangeable instruments, Computer Motion merged with Intuitive Surgical. Since then, there have been several significant improvements and many new generations of the platform, allowing adoption by multiple surgical specialties. Today, there are more than 5000 da Vinci systems around the world and over 6 million procedures performed, with general surgical procedures being the most widespread application [1]. As the United States performs 70.6% of robotic surgeries globally, the majority of this discussion will involve robotic surgery in the United States [2].

In the United States, more than 600,000 colon surgeries are performed annually [3]. Colorectal cancer is the third most common cancer in men and women, the third most common cancer cause of death, and the most common cancer cause of death for men under the age of 50 [4]. For patients with inflammatory bowel disease, 80% of patients with Crohn's disease and 30% of ulcerative colitis will require surgery in their lifetime [5]. As colorectal surgeries involve narrow working spaces like the bony pelvis and intricate dissections around vital structures, robotic surgery has been increasing utilized in order to apply advancing technology while maintaining a minimally invasive approach with its proven patient benefits [6].

2. Patient outcomes

The expanding adoption of robotic surgery in the colorectal field has led to increased research into patient outcomes. Many recent studies focusing on short-term outcomes of robotic surgeries are similar or even superior compared to open or laparoscopic approaches. The Robotic versus Laparoscopic Resection for Rectal Cancer (ROLARR) study was the largest international randomized control trial that compared the safety, efficacy, and short- and long-term outcomes of robotic colorectal procedures compared with conventional laparoscopy, and found no statistically significant difference in conversion rates to open laparotomy, circumferential resection margins, mesorectal resection quality, postoperative complications within 30 days, and quality of life on month 6 after operation. Meta-analyses from cohort and case-control studies on the use of robotics in rectal cancer resection suggests benefits in terms of all-cause mortality, surgical site infection, intraoperative blood loss, time to oral diet, postoperative return of gastrointestinal function, preservation of a functional anus, and urinary and sexual function [7].

Urologic and sexual function after surgery is an important area unique to colorectal surgery and not addressed in most studies. Up to 80% of men and 60% of women experience voiding and sexual dysfunction after rectal surgery. Some studies have shown faster recovery to normal urogenital function in robotic compared to laparoscopic groups, however, the ROLARR trial and other large-scale retrospective studies found no significant difference [6].

The benefits of robotic surgery for outcomes after cancer surgery are nuanced. When compared to laparoscopic surgery, robotic surgery has similar margin positivity rates, disease free survival, and overall survival for colon cancer. Robotic surgery has been shown to increase the number of lymph nodes harvested for all areas of the colon and rectum as well as improve radial resection margins in rectal cancer [7]. Conversely, advantages in this patient population were demonstrated in short-term outcomes such as shorter hospital stay and faster return of bowel function [8].

A proposed reason behind these benefits in short-term outcomes is related to decreased post-operative pain. A randomized study by Tolstrup et al. [9] comparing post-operative pain after robotic versus laparoscopic rectal cancer resection found that opioid consumption was significantly lower after robotic surgery compared to the laparoscopic group [10].

While there are many studies that demonstrate patient outcomes with robotic surgery, fewer studies explore patient perceptions and satisfaction. Existing literature reviews have found the majority of patient experiences were positive and patients were satisfied overall. Many perceived the new technology as better, felt that robotic surgery indicated a more technically skilled surgeon, and would recommend robotic surgery to others. Patients who were dissatisfied felt their length of stay shorter than expected [9, 11–13]. Patients were also more likely to be satisfied and comfortable with a robotic approach if they received preoperative education on the role of the robot during surgery and the maintained participation of the surgeon during their procedure [11, 12]. As patient satisfaction becomes increasingly important for hospital economics and reimbursements, setting appropriate expectations and educating patients on the role of technology during their operation is integral to improve this quality metric.

3. Surgeon outcomes

In a survey of almost 4000 surgeons by the American College of Surgeons in 2020, 13% of surgeons under 60 years of age were considering leaving surgery with only a 51% job satisfaction rate among them [14]. The top three reasons for leaving surgery were overall stress (79%), work time demands (77%), and personal time requirements (73%). With increasing life expectancy of the twenty-first century, there is a predicted shortage of almost 30,000 surgeons by the year 2030 [15]. Attenuating the loss of the surgical work force now will have downstream effects on impending physician shortages. Robotic surgery has the potential to improve job performance for surgeons by decreasing necessary personnel, improving ergonomics with a decrease in hand tremor, and options for decrease in musculoskeletal sequelae of traditional open and laparoscopic methods.

Robotic assisted surgery (RAS) has been shown to have higher task complexity for surgeons when compared to surgical techs and circulating nurses, a potential function of the physical space created by the surgeon console [16]. Aspects such as closed loop communication and situational awareness require more effort when the surgeon is sitting at the console with their head positioned into the eyepieces speaking through a microphone as opposed to standing a few feet away from their surgical scrub. Surgeon control of instruments from the console allows fewer personnel at bedside, potentially lowering healthcare costs in the face of healthcare worker shortage. On the Da Vinci Xi®, the surgeon essentially doubles his working arms from the traditional two used in laparoscopy to four. There is no longer the need for an assistant to hold the camera or hold a retraction arm as the surgeon, who is the most highly trained person in the operation, can control both the camera and retraction arm while maintaining control of two working arms [8].

Another improvement in surgeon efficiency during robotic versus open and laparoscopic surgery are the ergonomics. The console allows for adjustment tailored to an individual's body, leaving the variance to wrist motion, which is essentially constant throughout variations in human proportions, as opposed to hand size and height, which vary by surgeon. Physiologic degrees of freedom in the wrist including flexion-extension, radioulnar deviation, and rotation are mimicked in robotic instruments and even improved. When compared to the natural range of motion on the surgeon, the robotic instruments have 7 degrees of freedom and allow for 180 degrees of articulation and 540 degrees of rotation [8]. When observing participants with no surgical training, Prasad et al. found they were 2.5 and 7 times more accurate in moderate and fine motor skills, respectively, when compared to laparoscopic performance ($P < 0.001$) [17]. They also noted an equalized performance for dominant and non-dominant hands. In laparoscopic and open surgery, longer instruments act as a fulcrum between a surgeon's hand and the operative field, which serves to amplify tremor. The robot console eliminates this tremor, potentially an explanation for improved functional outcomes after pelvic surgery, such as sexual and bladder function [8].

Work-related musculoskeletal disorders range from 60 to 90% of surgeons based on specialty [18]. Pain from surgical practice also plays a role in less recognized phenomena such as quality of life and burnout which can trigger early retirement and surgeon shortages [19–21]. In general, it is recommended to maintain neutral positions of the neck and spine in both open and minimally invasive surgery. However, according to Stucky et al., laparoscopic surgeons are three to five times more likely to experience neck and shoulder pain than surgeons performing open procedures [22]. Using instruments that require less movement of the wrist in its physiologic degrees of freedom are recommended, leading to more time spent in awkward positions to accomplish challenging operative tasks. There are also limitations for surgeons with smaller hand sizes as most instruments are designed for glove sizes 6.5 and above and require the table to be lowered to pubic bone level as opposed to elbow level as in open surgery [18]. With the adjustable console and handpieces, these differences in size are eliminated. The height discrepancies between surgical team members to maintain an angle no greater than 30 degrees between eye level and the laparoscopic monitor for favorable neck positioning is also eliminated by having the surgeon at the robot console [23]. Hokenstad et al. have developed an extensive guide to setting up the surgeon console for optimal positioning from angle of neck and back flexion to differences in forehead placement to avoid headaches [24]. Although guides such as this exist, there is concern that training residents and medical students are not exposed to the importance of ergonomics and solutions for "awkward" postures. In a survey sent by the ACGME in 2019, only 1.5% of program directors reported a formal ergonomics curriculum, serving as a missed opportunity to improve potentially detrimental behaviors early in careers and avoid lifelong musculoskeletal pain and early retirement [25].

4. Training/learning curve

The American Board of Surgery, which is the accrediting body responsible for certifying surgeons in the United States, recommends that all trainees have knowledge of robotic surgery, however there is no standardized robotics curriculum and no requirement for robotic cases exists for board eligibility [26]. Although the Board has added the option on the case logging system to declare a case as robotic, the CPT code used for billing and tracking is the same as its laparoscopic counterpart. With the extensive simulation training options and improved ergonomics of robotic surgery, there is a reported shorter learning curve. Studies report a curve of 30–100 cases for laparoscopic surgery vs. 15–44 cases for robotic surgery [8]. For rectal surgery specifically, operative time to perform a total mesorectal excision was significantly reduced within the first 20 robotic cases for surgeons who previously performed the procedure open [27].

In a review of experienced surgeon opinion as they observed robotic case content, Green et al. noted that robotic surgery depends more on visual assessment. This is due to the lack of tactile feedback with an emphasis on risk avoidance [28]. When teaching residents, their surgeons employed strategies such as dual console and simulation with communication of visual information to be the mainstay of teaching. Cope et al. comment on this as a surgical "sensory semiosis" constructed from visual and haptic clues and replacing traditional tactile feedback [29]. While the loss of tactile feedback is often listed as a weakness of the robotic approach, early robotic assisted surgery models developed MIT and Stanford under the original Department of Defense and NASA grant found that surgeons performed with higher accuracy using 3D visualization as opposed to haptic feedback available on earlier robotic prototypes [1]. Despite the differences in sensory feedback between robotic and open surgery, surgeon discussion emphasized the importance of appropriate exposure and tension, themes common to traditional open and laparoscopic surgery as well.

As previously stated, risk avoidance is particularly important in the surgical teaching environment. In order to standardize an approach specific to robotic surgery, Kalipershad and Peristerakis developed a safety protocol with an emphasis on clear communication, situational awareness and role clarity, with the goal of reducing errors in a high-pressured environment. The protocol has two main areas of concentration, the first of which being on human factor simulation training with operative team members. The second area is a step by step plan for "emergency undocking" which would be triggered by the operative surgeon and lead to a cascade of events for patient safety and efficient conversion to open [30].

An interesting trend in training for RAS is the discrepancies in teaching by gender. In a study by Foley et al. comparing console participation time per case, they found that male attending surgeons provided female trainees less console participation than male counterparts (52.1% vs. 59.7%) while female attending surgeons provided same amount of time [31]. When looking specifically at total mesorectal excision, male trainees performed more complete procedures than their female counterparts (57% vs. 42%) [31]. By allowing objective data collection, robotic surgery can also be a tool to support equitable training environments for an increasingly diverse surgical work force.

Simulation is an important part of robotic surgery training for not only residents and medical students but practicing attendings who want to adopt robotic assisted surgery into their practice. While systematic reviews have identified the value of simulation and virtual reality training in laparoscopic skill improvement, little has been proven in robotic skills [32–34]. Available simulation models include the Fundamentals of Robotic Surgery physical Dome, the da Vinci Simulation System (DVSS), and the dV-Trainer. When comparing performance on these training devices to a control group, all groups significantly improved their performance [35]. Some centers are recommending a "virtual reality warm-up" prior to starting robotic cases [36, 37]. When testing both residents and experienced MIS surgeons improvements were shown in task time, cognitive errors, and intra-corporeal suturing after a 3–5 minute simulator warm up [37].

A concern for the decreased need of bedside assistance during robotic surgery is the loss of medical student involvement. Stimulating medical student interest in the field of surgery is another mechanism to combat the impending surgeon shortage. Specific concerns involve physical separation from the operating attending/resident affecting engagement with the case and the monotony of "just watching" instead of physically participating in the surgery. When surveying fourth year medical students at the Medical College of Wisconsin, Higgins and O'Sullivan found that the visual and audio observation actually enhanced their learning. They were able to hear surgeons more effectively with the microphone feature of the robot and they could follow the case progression on an adjustable screen. Unfortunately, the medical students in the study generally found the robotic surgery learning environment to be demotivating [38]. It is hard to separate if these feelings are specific to the robotic surgery learning environment or generalized recognized themes among medical students such as the lack of self-determination, with no ability to develop their autonomy, motivation through skill competence, or regular communication with the attending surgeon [39].

5. Economic considerations

The introduction of new technology does introduce new costs. Although robotic surgery has existed since 2000, for much of this time period the Intuitive DaVinci system has been the only FDA approved platform in the United States [26]. Although Intuitive has continued to develop new instruments with an increased number of uses and a lease type arrangement, the initial startup cost of a robotic surgery system can still be cost prohibitive for smaller hospitals or low- and middle-income countries [2]. Although Intuitive has maintained substantial market share, the patents for the DaVinci system expired in 2016 which should introduce more competition with a subsequent decrease in cost [40]. When compared to laparoscopic surgery, robotic surgery does have an increased cost for the strict operating room costs for performing an operation by approximately $1–3 thousand dollars [41]. However, healthcare costs are not strictly operating room costs. While many operations are attempted in a minimally invasive fashion, whether laparoscopic or robotic, many operations are converted to a larger open laparotomy incision for a variety of reasons. The conversion to an open surgery is significantly lower for robotic than laparoscopic surgery with a shorter length of inpatient stay [42]. When comparing the cost of a robotic colorectal surgery operation to a laparoscopic surgery converted to an open operation, the cost of the robotic surgery is lower [6]. Across all payers in the United States, 1 day of an inpatient stay costs approximately $1000 [43]. Shorter lengths of stay would increase a hospital system's ability to admit more patients and generate higher revenues for the system as a whole as well as offset the increase in the original cost of a robotic operation versus a laparoscopic operation. Early discharges, defined as less than or equal to 3 days after a minimally invasive operation are also associated with a decreased risk of post-discharge complications which are another driver of overall increased healthcare expenditure [44]. Hospital cost increases from surgical complications can range anywhere from $3000 to over $100,000 depending on the severity and nature of the complication [45].

As many robotic instruments have a finite number of uses, unlike open instruments which can mostly be sterilized and reused indefinitely, accuracy of documentation and charge capture is increasingly important for cost containment. In newly established robotic programs, error rates of 60% when logging robotic items can occur. By implementing improved peer education, barcode systems, and improved item descriptions, error rates can be decreased by 75% which increases the ability to bill for robotic items. The Intuitive system also directly captures which instruments have been utilized. Although these logs cannot be used directly for billing purposes, they can be compared to operating room documentation as an objective measure of team accuracy and performance [46].

A major limitation of the literature reviewing costs of robotic surgery is variability in the surgeon learning curve. For the treatment of colorectal cancer, robotic technology has been preferentially adopted by high inpatient volume hospitals, teaching hospitals, and has had a quicker rise for treatment of rectal cancer than colon cancer [47]. Higher utilization of robotic surgery is also associated with a higher number of general surgeons, higher hospital density within a service area, higher average income, and a lower proportion of uninsured patients [48]. A major critique of RAS is the increased operating time, which is associated with increased length of stay and more frequent complications in colorectal surgery patients, regardless of the surgical approach [49]. However, when evaluating fellowship trained colorectal surgeons from 2015 to 2019, Johns et al. found that the total cost of hospitalization was 25% greater for the laparoscopic group and the robotic surgeries were shorter in duration than the laparoscopic operations [41]. Looking specifically at abdominoperineal resection for rectal cancer, which involves removing the entire rectum and anus, Gorgun et al. found no difference in overall costs and decreased operative times for the robotic surgeries compared to the laparoscopic surgeries [50].

Cost containment can also occur at the operating room management level by addressing staffing assignments. When applying the NASA task load index to staff within an operating room during a robotic operation, Walters et al. [51] identified that key steps of the operation required highly time dependent effort and physical demands. These key times included room setup, start of the operation, patient preparation for surgery, registered nurse documentation and management of surgical specimens, moving patients at the end of the procedure, and room cleanup. In order to minimize the demands on the more highly paid members of the team, the group recommended that registered nurses do not participate in bedside scrubbing for procedures and identified opportunities to engage non-licensed support staff. Effective use of non-licensed support staff enables more cost-effective staffing models [51].

Changes in disease processes and outcomes also have downstream effects on healthcare costs and patient income. The most common disease processes of the colon and rectum are associated with obesity, a condition which has been steadily increasing throughout the United States [52]. Obesity introduces a new set of surgical challenges, including increased conversion to open and diminished volume of space to operate in a minimally invasive fashion. Robotic surgery has been preferentially applied to pelvic surgery in obese men and been associated with decreased conversion to open, decreased sexual and autonomic dysfunction, and early decrease in associated urinary symptoms [53]. Other oncologic outcomes that are improved in robotic surgery include higher lymph node harvests [7] and improved radial resection margins in rectal cancer [54]. For patients with a complete tumor resection but a lymph node harvest lower than 12, many recommendations include adjuvant chemotherapy, which adds costs for port placement and administration of chemotherapy medications, which often include intravenous infusions, monitoring, and frequent laboratory evaluations. Additionally, male obese patients historically have higher positive radial margins and lower lymph node harvests for rectal cancer. The technological improvements in robotic surgery help mitigate those differences and save downstream costs for this vulnerable patient population [55].

The most common indications for colon resection, diverticulosis and colon cancer, are also occurring in increasingly younger patient populations [56]. Disease processes that typically occurred in ages 50 or above have shifted to those younger than 50. In this population, patients are more likely to be building their careers and have dependent children under the age of 18 living at home [57]. Disruptions in career and potential ability to support a dependent family are more of a personal financial burden for these patients, a cost and quality metric which is not often accounted for in healthcare cost literature. Robotic surgery is not only associated with a shorter length of stay in the hospital but is also associated with a faster return to work by approximately 4 days [58]. This quicker return to work mitigates individual economic stressors for patients.

Another currently theoretical benefit of the improved ergonomics of robotic surgery is career longevity of surgeons. As previously mentioned, the United States is predicted to have a deficit of 23,000 general surgeons by the year 2032. The shortage is due to multiple effects such as surgeon retirement during the COVID-19 pandemic and decreased interest in surgery by medical school graduates [59]. Resident training is also funded by Medicare, which spent $4.5 billion in 2020 for resident education [60]. As physician education is a significant usage of taxpayer dollars, improvements in surgeon longevity not only addresses the impending physician shortage but also improves the return on investment of government spending on physician training.

6. Future directions

Despite the lack of long-term outcomes data, robotic surgery continues to grow in the field of colorectal surgery. The FDA has recently approved the Senhance and Versius systems which will introduce competition into the market, as well as new configurations and features that may offset some of the critiques of the Intuitive systems, such as a closed design that separates the surgeon from the operating team. Although remote surgery was the original intention by the Department of Defense, current remote applications are limited to proctoring which can improve healthcare equity by reducing barriers to training and mentorship. Application of robotic assisted surgery to endoscopy may enable lower insufflation pressures with lower risk of perforation as well as the ability to remove larger tumors without the need for a formal operation. Integration of artificial intelligence (AI) may also enable new tools to improve operative times and accuracy. 3D rendering of CT scan findings may also be rendered onto operative images to facilitate identification of pathology [10].

DOI: http://dx.doi.org/10.5772/intechopen.1004548

7. Conclusion

Despite criticism of cost and lack of long-term data, robotic assisted surgery has become an integral part of colorectal surgery. From a health outcomes side, robotic colorectal surgery is associated with decreased pain, decreased length of stay, decreased risk of surgical site infections, increased lymph node harvest for cancer, improved resection margins for obese men with rectal cancer, improved urinary and sexual function, and an expedited return to work. For surgeons, robotic surgery allows for improved ergonomics accommodating a wider variety of body types and potentially a longer career. Robotic systems also enable objective collection of data during surgeries that facilitates systematic approaches to managing necessary services with fewer resources. For surgeons who are less experienced with minimally invasive surgery, robotics allows for quicker adaptation of a minimally invasive approach with fewer conversions to open. For surgeons with fellowship training, robotic colorectal surgery can be performed faster than laparoscopic surgery with lower overall hospitalization costs. When used correctly, robotic assisted surgery adds value for patients, hospital systems, and surgeons.

Author details

Alexandra Green, Ashley Alexander and Bethany Malone*
Texas Health Fort Worth, Fort Worth, TX, United States

*Address all correspondence to: bcbmalone@gmail.com

References

[1] Ranev D, Teixeira J. History of computer-assisted surgery. Surgical Clinics of North America. 2020;**100**:209-218

[2] Mehta A, Ng JC, Awuah WA, et al. Embracing robotic surgery in low- and middle-income countries: Potential benefits, challenges, and scope in the future. Annals of Medicine and Surgery. 2022;**84**:104803

[3] Colon Resection Surgery. Society of American Gastrointestinal and Endoscopic Surgeons (SAGES). Web. 2022

[4] US Cancer Statistics Working Group. US Cancer Statistics Data Visualization Tool, based on 2022 Submission Data (1999-2020): US Department of Health and Human Services, Centers for Disease Control and Prevention and National Cancer Institute. Available from: https://www.cdc.gov/cancer/dataviz [Accessed: November 2023]

[5] Schwartzberg D, Remzi F. The role of laparoscopic, robotic, and open surgery in uncomplicated and complicated inflammatory bowel disease. Gastrointestinal Endoscopy Clinics of North America. 2019;**29**:563-576

[6] Baek S, Piozzi GN, Kim S. Optimizing outcomes of colorectal cancer surgery with robotic platforms. Surgical Oncology. 2022;**43**:101786

[7] Yuval JB, Thompson HM, Verheij FS, et al. Comparison of robotic, laparoscopic, and open resections of nonmetastatic colon cancer. Diseases of the Colon and Rectum. 2023;**66**:1347-1358

[8] Shah MF, Nasir II, Parvaiz A. Updates in robotic colorectal surgery. Surgery. 2019;**38**(1):38-42

[9] Tolstrup R, Funder JA, Lundbech L, Thomassen N, Iversen LH. Perioperative pain after robot-assisted versus laparoscopic rectal resection. International Journal of Colorectal Disease. 2018;**33**(3):285

[10] Erozkan K, Gorgun E. Robotic colorectal surgery and future directions. The American Journal of Surgery [Web]. 26 Oct 2023

[11] Moloney R, O'Brien B, Coffey JC, et al. Patients' perceptions after robot-assisted surgery: An integrative review. AORN. 2020;**112**(2):133-141

[12] Anand A, Mahmood Z, Dixon F, et al. Patient perceptions of and postoperative satisfaction with colorectal robotic surgery: A qualitative insight. European Journal of Surgical Oncology. 2024;**50**:107141

[13] Moore NB, Bhatti SS, Raskin ER, et al. Patient perceptions on robotic assisted surgery: A systematic needs assessment. Journal of the American College of Surgeons. 2023;**236**:S102-S103

[14] Mahoney ST, Strassle PD, Schroen AT, et al. Survey of the US surgeon workforce: Practice characteristics, job satisfaction, and reasons for leaving surgery. Journal of the American College of Surgeons. 2020;**230**(3):283-293.e1

[15] Williams TE Jr, Satiani B, Thomas A, Ellison CE. The impending shortage and the estimated cost of training the future surgical workforce. Annals of Surgery. 2009;**250**:590-597

[16] Zamudio J, Woodward J, Kanji FF, et al. Demands of surgical teams in robotic-assisted surgery: An assessment of intraoperative workload

within different surgical specialties. The American Journal of Surgery. 2023;**226**:365-370

[17] Prasad SM, Prasad SM, Maniar HS, et al. Surgical robotics: Impact of motion scaling on task performance. Journal of the American College of Surgeons. 2004;**199**:863-868

[18] Tetteh E, Wang T, Kim JY, et al. Optimizing ergonomics during open, laparoscopic, and robotic-assisted surgery: A review of surgical ergonomics literature and development of educational illustrations. The American Journal of Surgery [Web]. 10 Nov 2023

[19] Howarth AL, Hallbeck MS, Lemaine V, Singh DJ, Noland SS. Work-related musculoskeletal discomfort and injury in craniofacial and maxillofacial surgeons. The Journal of Craniofacial Surgery. 2019;**30**(7):1982-1985

[20] Davila VJ, Meltzer AJ, Hallbeck MS, Stone WM, Money SR. Physical discomfort, professional satisfaction, and burnout in vascular surgeons. Journal of Vascular Surgery. 2019;**70**(3):913-920

[21] Wells AC, Kjellman M, Harper SJF, Forsman M, Hallbeck MS. Operating hurts: A study of EAES surgeons. Surgical Endoscopy. 2019;**33**(3):933-940

[22] Stucky CCH, Cromwell KD, Voss RK, et al. Surgeon symptoms, strain, and selections: Systematic review and meta-analysis of surgical ergonomics. Annals of Medicine and Surgery. 2018;**27**:1-8

[23] Hill SG, Kroemer KHE. Preferred declination of the line of sight. Human Factors. 1986;**28**(2):127-134

[24] Hokenstad ED, Hallbeck MS, Lowndes BR, et al. Ergonomic robotic console configuration in gynecologic surgery: An interventional study. Journal of Minimally Invasive Gynecology. 2021;**28**(4):850-859

[25] Epstein S, Tran BN, Capone AC, et al. The current state of surgical ergonomics education in U.S. surgical training: A survey study. Annals of Surgery. 2019;**269**(4):778-784

[26] Arca MJ, Adams RB, Angelos P, et al. American board of surgery statement on assessment and robotic surgery. The American Journal of Surgery. 2021;**221**:424-426

[27] Pigazzi A, Luca F, Patriti A, Valvo M, Ceccarelli G, Casciola L, et al. Multicentric study on robotic tumor-specific mesorectal excision for the treatment of rectal cancer. Annals of Surgical Oncology. 2010;**17**(6):1614-1620

[28] Green CA, Lin J, Higgins R, et al. Expertise in perception during robotic surgery (ExPeRtS): What we see and what we say. The American Journal of Surgery. 2022;**224**:908-913

[29] Cope AC, Mavroveli S, Bezemer J, Hanna GB, Kneebone R. Making meaning from sensory cues: A qualitative investigation of postgraduate learning in the operating room. Academic Medicine. 2015;**90**(8):1125-1131

[30] Kalipersha SNR. Peristerakis. The introduction of an emergency safety protocol coupled with simulation training in robotic surgery, has enabled a more cohesive and efficient response to emergencies. The Surgeon. 2022;**20**:151-156

[31] Foley KE, Izquierdo KM, von Muchow MG, et al. Colon and rectal surgery robotic training programs: An evaluation of gender disparities.

Diseases of the Colon and Rectum. 2020;**62**:974-979

[32] Larsen CR, Oestergaard J, Ottesen BS, et al. The efficacy of virtual reality simulation training in laparoscopy: A systematic review of randomized trials. Acta Obstetricia et Gynecologica Scandinavica. 2012;**91**:1015-1028

[33] Nagendran M, Gurusamy KS, Aggarwal R, et al. Virtual reality training for surgical trainees in laparoscopic surgery. Cochrane Database of Systematic Reviews. 2013:CD006575

[34] Nagendran M, Toon CD, Davidson BR, et al. Laparoscopic surgical box model training for surgical trainees with no prior laparoscopic experience. Cochrane Database of Systematic Reviews. 2014:CD010479

[35] Satava RM, Stefanidis D, Levy JS, et al. Proving the effectiveness of the fundamentals of robotic surgery (FRS) skills curriculum: A single-blinded, multispecialty, multi-institutional randomized control trial. Annals of Surgery. 2020;**272**:384-392

[36] Kelly JD, Kowalewski TM, Brand T, et al. Virtual reality warm-up before robot-assisted surgery: A randomized controlled trial. Journal of Surgical Research. 2021;**264**:107-116

[37] Lendvay TS, Brand TC, White L, et al. Virtual reality robotic surgery warm-up improves task performance in a dry laboratory environment: A prospective randomized controlled study. Journal of the American College of Surgeons. 2013;**216**:1181-1192

[38] Higgins RM, O'Sullivan P. The robotic surgery learning experience through the eyes of the medical student: What do they see? Journal of Surgical Education. 2019;**77**:549-556

[39] Ten Cate OTJ, Kusurkar RA, Williams GC. How self-determination theory can assist our understanding of the teaching and learning processes in medical education. AMEE Guide No. 59. Medical Teacher. 2011;**33**:961-973

[40] Maitra I, Date RS. Robotic surgery: Is the technological advance worth the bravado? International Journal of Surgery. 2019;**62**:1-2

[41] Johns MS, Gough B, McCarthy E, et al. Cost comparison of laparoscopic and robotic colon resections at a hybrid academic/community institution. Journal of the American College of Surgeons. 2020;**231**:e94

[42] Addison P, Agnew JL, Martz J. Robotic colorectal surgery. Surgical Clinics of North America. 2020;**100**:337-360

[43] Moore B, Levit K, Elixhauser A. Costs for Hospital Stays in the United States, 2012. HCUP Statistical Brief #181. Rockville, MD: Agency for Healthcare Research and Quality; 2014. Available from: http://www.hcup-us.ahrq.gov/reports/statbriefs/sb181-Hospital-Costs-United-States-2012.pdf

[44] Balas M, Quereshy F, Bohnen J, et al. Early discharge after uncomplicated elective colectomy and risk of post-discharge complication. Journal of the American College of Surgeons. 2024;**238**:182-196

[45] Ludbrook GL. The hidden pandemic: The cost of postoperative complications. Current Anesthesiology Reports. 2022;**12**:1-9

[46] Gerull WD, Pierce A, Mody J, et al. Capturing and improving case charge accuracy in robotic surgery programs. Journal of the American College of Surgeons. 2022;**234**:964-968

[47] Schootman M, Hendren S, Ratnapradipa K, et al. Adoption of robotic technology for treating colorectal cancer. Diseases of the Colon and Rectum. 2016;**59**:1011-1018

[48] Fieber JH, Wirtalla C, Kuo LE, et al. Variation in the utilization of robotic surgical operations. Diseases of the Colon and Rectum. 2018;**227**:e145

[49] Unruh KR, Bastawrous AL, Kanneganti S, et al. The impact of prolonged operative time associated with minimally invasive colorectal surgery: A report from the surgical care outcomes assessment program. Diseases of the Colon and Rectum. 2024;**67**:302-312

[50] Gorgun E, Cengiz TB, Ozgur I, et al. Outcomes and cost analysis of robotic versus laparoscopic abdominoperineal resection for rectal cancer: A case-matched study. Diseases of the Colon and Rectum. 2022;**65**:1279-1286

[51] Walters C, Webb PJ. Maximizing efficiency and reducing robotic surgery costs using the NASA task load index. AORN Journal. 2017;**106**:283-294

[52] Adult Obesity Prevalence Maps. Centers for Disease Control and Prevention. National Center for Chronic Disease Prevention and Health Promotion, Division of Nutrition, Physical Activity, and Obesity. 2023

[53] Crain N, Ho NJ, Aboulian A. A comparative analysis of short-term patient outcomes after laparoscopic versus robotic rectal surgery. Diseases of the Colon and Rectum. 2022;**65**:1274-1278

[54] Gachabayov M, Bergamaschi RC. Outcomes of robotic surgery in obese males with low rectal cancer. Diseases of the Colon and Rectum. 2020;**231**:e105

[55] Flynn J, Larach JT, Kong JCH, et al. Patient-related functional outcomes after robotic-assisted rectal surgery compared with a laparoscopic approach: A systematic review and meta-analysis. Diseases of the Colon and Rectum. 2022;**65**:1191-1204

[56] Saraiva MR, Rosa I, Claro I. Early-onset colorectal cancer: A review of current knowledge. World Journal of Gastroenterology. 2023;**29**:1289-1303

[57] Siegel RL, Miller KD, Sauer AG, et al. Colorectal cancer statistics. CA: A Cancer Journal for Clinicians. 2020;**70**(2020):145-164

[58] Simianu VV, Gaertner WB, Kuntz K, et al. Cost-effectiveness evaluation of laparoscopic vs robotic minimally invasive colectomy. Annals of Surgery. 2020;**272**:334-341

[59] Haskins J. Desperately seeking surgeons. AAMC News. 2019. Available from: https://www.aamc.org/news/desperately-seeking-surgeons

[60] Medicare graduate medical education payments: An overview. Congressional Research Service Reports. 2022. Available from: https://crsreports.congress.gov/product/pdf/IF/IF10960